TAKE CAR OF
YOUR
AND LE

"Take care andle,
be charit poor
and pray dead."

old Night Watch Curfew cry

An account of eleven disastrous fires in small Devonshire towns during the Nineteenth Century

by F. D. Gentry

DEVON BOOKS

First published 1985 by Devon Books

Gentry, F. D.
Take care of your fire and candle: an account of eleven disastrous fires in small Devonshire towns during the nineteenth century.
1. Fires—England—Devon—History—19th century
2. Devon—History
I. Title
942.3′5 DA670.D5
ISBN 0-86114-759-6

ISBN 0 86114 759 6

DEVON BOOKS

Publishers to Devon County Council

Devon Books is a division of A. Wheaton & Co Ltd, which represents a consortium of three companies:

Editorial & Design
Limelight Ltd, 15 Castle Street, Exeter EX4 3PT

Production & Manufacturing
A. Wheaton & Co Ltd, Hennock Road, Exeter EX2 8RP
(A. Wheaton & Co Ltd is a division of Pergamon Press)

Sales & Distribution
Town & Country Books, 24 Priory Avenue, Kingskerswell, Newton Abbot TQ12 5AQ

Printed in Great Britain by A. Wheaton & Co Ltd

DEDICATION

To the memory of my late wife,
AMY FRANCES

Winter has come yet still the roses bloom
Fragrant beneath the window of our room,
Their scented breath your valediction sighs
To melt the ice-cold tears that scald my eyes,
For Summer fled and Autumn took to wing,
Winter has come—but there can be no Spring.

F.D.G.

CONTENTS

ACKNOWLEDGEMENTS

It is with sincere gratitude that I thank the following friends for their invaluable help: Oswald and Phyllis Moss, who devoted countless hours to the task of unearthing newspaper reports and recording them on tapes for my use; Jack Berry for his patient proof-reading and invaluable advice; Delia Noble and Eileen Davey for the hours they spent reading obscure reports from indistinct photocopies or bringing order to a chaos of taped notes. My thanks to all of the above for their constant support and encouragement.

INTRODUCTION

The purpose of this book is to take a critical, if sympathetic, look at life among the inhabitants of some small Devonshire towns and villages ravaged by fire in the 19th century and to observe, through newspaper and other reports, how people from all walks of life reacted, not only to the immediate terrifying experience but also towards one another.

The towns dealt with in this study were chosen either because a single devastating fire all but destroyed them or, as in the case of South Molton and Cullompton, for a number of serious outbreaks which occurred throughout the century and, in other instances, because newspaper accounts of the fires contain details which throw light upon certain interesting aspects of prevailing social attitudes and conditions.

The changes which have occurred since the early 1800s—political, economic, social and technological—may have been so far-reaching that many people of the present century find it difficult to understand the attitudes and behaviour of their forebears as they went about their daily lives. There is an abundance of literature, in the form of learned histories and novels, providing a variety of opinions on the relationships between different social sections of English country society during the development of the Industrial Revolution. In many instances such works concentrate on the exploitation of the labouring class by capitalist employers. This is, however, a misleading over-simplification which offers little insight into the human relationships and the undertow of reasons prompting them. There can be no doubt that exploitation of one kind or another went on in the factories and mills, as well as in some of the great houses of the aristocracy where a domestic servant could rate rather lower than a gun dog.

One aspect of the newly developing industries, which the enquirer soon observes, is the change brought to the workers' prospects. The labourer on the land had, for centuries, received little more than subsistence so that he, and his family, were virtually tied to the farm with no prospect of any alteration in the pattern of their lives. In Devonshire, however, many people were engaged in the production of woollen serge, either in mills or in their homes, and these, like workers in the new industries, received cash payments so that the thrifty and the ambitious could actually save money which could eventually bring them independence. After twenty years of regular employment a man may have saved enough to buy a few acres of land, build himself a dwelling and produce enough to feed his family.

By modern standards the factory wages were deplorably low in comparison with the large profits the owners made but moral awareness of such differences had not developed in the social conscience and it is doubtful if labourers in the early 19th century felt any great sense of being wronged while employers saw no harm in paying the going rate for the job, however low. Such was the outlook of employer and worker in the small towns of Devonshire as elsewhere at this period.

Ever since man contrived to take fire into his dwelling he has been paying the price for the convenience in terms of danger to life as well as property. Stone or brick fireplaces and chimneys were introduced into wooden and cob houses in an attempt to minimise the danger from misfortune or sheer carelessness which to this day still threaten every building to a greater or lesser degree depending upon design and construction materials.

In the early 19th century people seem to have taken a fatalistic attitude towards fire, almost as if they believed its powers were magical and man's ingenuity incapable of devising anything to control the element. In earlier centuries the fire on the hearth, as well as rushlights and candles, had to be extinguished when the Curfew bell was rung at night-fall, so that although the sleeper might suffer from the cold in winter, he was less liable to be roasted in his bed.

Exeter was among the first provincial cities in England to have a fire-engine of any sort. The contraption was purchased in 1626 and was entirely manual. It was carried to the scene of the fire by two men to whose shoulders its supporting leather straps were fastened and could only have been effective when dealing with the smallest outbreaks. A container had to be kept filled with water which was forced by a sort of plunger through a pipe made of leather. Leather buckets, used for fire-fighting since Roman times, were probably more effective so long as there were enough of them and also plenty of water to hand.

The first mobile water-pumping fire-engine, a four-wheeled vehicle, was introduced in London by the Royal Exchange Insurance Company in the early 18th century, and in Devonshire the town of South Molton has records showing that it bought its first fire-engine as early as 1736. Exeter's first machine of this kind was purchased by the West of England Fire Insurance Company in 1807, and became affectionately known to the inhabitants as 'The Little West'. This remarkable machine did good service in and around the City of Exeter for some eighty years. Other insurance companies followed the West of England's lead and stories are told of brigades refusing to tackle fires on premises insured by their rivals. Each company affixed a distinctive plaque in a conspicuous place to property they insured. Some of these may still be seen on old buildings but they are disappearing rapidly since becoming what are called 'collectors' items'. The rival companies soon realized, however, the necessity of coming to an amicable arrangement with each other to share costs and to assist one another in fighting fires.

Since it is the intention of this book to scrutinise and attempt to understand the attitudes of the people involved in disastrous fires, it may be appropriate here to recall some of the traditional work the people of Devonshire did, and the conditions in which they lived and toiled.

From the 14th century more people in Devonshire worked in the production of woollen serge than worked on the land, and Exeter was by the 17th century the fourth largest cloth manufacturing town in England. At the beginning of the 18th century all the fields between Holloway Street and the River Exe were occupied by row upon row of frames with serge stretched on tenterhooks to dry and for a distance of twenty miles around the city people were engaged in spinning, weaving, dressing, fulling and dyeing. Paper also began to be manufactured on a large scale in Devonshire and in order to have labour near at hand small dwellings built of the cheapest material were packed together in narrow streets near the mills.

By the time the 19th century dawned, however, finer cloths from Holland, as well as from Lancashire, had all but killed the Devonshire serge industry and its manufacturers were only kept going by the East India Company which, until 1838, had the monopoly of trade with China, at whose ports they dumped large quantities of the serge at prices well below cost. It requires little imagination to enable us to guess that wages, in such circumstances, would have been at rock bottom and many cottage looms were all too often idle and most of the mills closed.

Due to the rapid decline of the cloth industry workers had to look for other employment near at hand. Enterprising individuals turned their hands to anything by which they might earn a living and, even in the smallest towns, one-man businesses such as soap boiling and candlemaking sprang up while the numbers employed in the traditional crafts, such as saddlers, tanners, coopers, wheelwrights, etc., increased.

In towns and villages fortunate enough to be located on important cross-roads, many people found gainful employment in the inns attending to the wants of travellers who thronged the roads. The people were industrious and shrewd and, despite a degree of poverty, even when judged by the meagre standards prevailing at the beginning of the 19th century, Devonshire was probably as good a place to live in as any other. It is difficult, if not impossible, to make any useful comparisons with present-day standards. The value of money in terms of purchasing power is particularly difficult to assess because in one part of the country a shilling might buy little more than half the quantity of any given commodity than it would have elsewhere and, even in places only half a day's journey apart, such differences could mean poverty here, yet modest plenty there. In his book, *The Age of Revolutions* (Mentor, New English Library) Professor Hobsbawm also points out that 'Indices of real wages cannot be drawn with any pretension to accuracy.'

Most of the towns dealt with here were once ancient riverside settlements and, therefore, suitable for Devon's staple industry—the manufacture and processing of woollen serge.

The houses in which most of the working people of the small Devonshire towns lived were built of cob, that is of local clay mixed with chopped straw and with stones added, and supported by a considerable amount of timber framing, and roofed with straw thatch which was warm in winter and cool in summer. It was picturesque and, above all, cheap. Hot, dry weather with a steady breeze blowing across the open moors, however, dried these materials to a point where a spark from kitchen, bakehouse or forge could set a roof ablaze. Tufts of flaming thatch would then be carried by the wind to ignite other buildings nearby until an entire town could become engulfed in fire.

The first mobile fire-engines were high-sided carts made to hold water in the centre of which were brass pumps. These were operated by half a dozen men who worked the long wooden detachable pump-handles. They depended upon a ready supply of water to be effective. All too often this was insufficient for the engine to be of much practical use and the fixed metal pipe through which the water was pumped made it necessary to place the machine close up to the burning building. Many were the occasions when a mass of blazing thatch would fall on it, set it on fire and destroy it. This was overcome later in the century when the flexible canvas hose with brass couplings was invented which also had the merit of making it possible to join long lengths of hoses so that large quantities of water, in rivers or lakes, could be brought into use.

The middle of the century saw the advent of the steam-pump mounted on the fire-engine and capable of throwing a powerful jet of water more than a hundred feet into the air. The next progressive step was to mechanize the fire-engine itself and steam-driven ones began to appear, but later the development of the internal combustion engine made steam obsolete.

In the early part of the 19th century the streets of most of the little Devonshire towns were exceedingly narrow and, when fire broke out, must have acted like horizontal chimneys, creating enough draught, even on a quiet day, to fan the flaming buildings into an inferno with frightening rapidity.

Because of this, and the inflammable nature of the building materials, the first thoughts of the occupants were not to try to extinguish the blaze but to make sure of their own safety, and that of their families, before salvaging what they could of their possessions. To this end they were prepared to smash the upstairs windows and throw out chests of drawers or linen chests into the street below. Sometimes they discovered, upon reaching the street, that their furniture was smashed to matchwood by the fall and that the contents had been looted by neighbours. Where there was a horse, pigs or poultry in the outbuildings behind the cottage these received immediate attention as they were usually their owner's most valuable possessions. In spite of the risks people were sometimes prepared to take to save livestock or chattels, it is remarkable how few fatalities occurred when fire swept through the country towns, even when up to two-thirds of the dwellings and shops were gutted in a few hours.

In the large towns of Plymouth and Exeter it was a different story. Almost all the houses within the city walls were of brick or stone with slated roofs and, although many fires occurred there was seldom any extensive spreading but, because some were built as high as four or five storeys residents on the upper floors sometimes perished, the staircases blazing freely in the draught caused by the stairwell acting like a chimney. Another difference between the large towns and the smaller ones was the plenitude of water usually available in the former, as well as fire-fighting appliances and trained brigades being always on call.

Two other points of importance to this study should here be mentioned. The first is the considerable difference in the matter of transportation between the first half of the 19th century as compared with the latter half. In the earlier years people had to move from place to place on foot, on horseback or by a horse-drawn vehicle; the carrier's cart for the poor, the stage-coach for the not so poor and the private carriage, chaise or curricle for the rich. By the middle of the century, railways were linking up all the large towns while branch lines connected the smaller ones and both goods and passengers were carried to their destinations at speeds of up to fifty miles an hour. A twenty mile journey before the railways came could take half a day as compared with as little as half an hour on a train.

The second point to remember about those times in Devonshire and all over England was that many artisans and labourers were drafted into the army and navy during the war with France, which had started in 1793 and continued with only two short breaks until 1815. After Waterloo, however, the return of so many men from the services, all seeking employment and homes to live in, would have raised considerable problems, and it was to the advantage of the little towns that the majority of these ex-servicemen tended to prefer the large urban areas in which to settle.

1

CHULMLEIGH

The summer of the year 1803 was a scorcher and the Devon countryside was parched. Wheatfields shimmered golden in the sunshine, swaying gracefully as the breeze swept the heavy heads into motion like gentle waves on a slack tide. The many streams and rivers were running perilously low and wells everywhere had completely dried up by the beginning of August.

In spite of its elevated position on one of those North Devon hills that rise like a congealed bubble and about which there is almost always a breeze blowing, the town of Chulmleigh must, nevertheless, have been stiflingly hot. Its narrow streets may well have provided a certain amount of shade, yet anyone entering their confines would have encountered such heat as to remind them of standing before an open oven door on baking day.

Heat or no heat, however, there was work to be done, for Chulmleigh was a busy market town standing astride busy cross-roads which brought a constant stream of carts, wagons and coaches with sweating horses stamping and struggling up the main street under a cloudless sky.

A year before, on 27th March 1802, the Peace of Amiens had been signed to end the war with France which had been in progress for nine years. The respite from war was short-lived and, on 18th May 1803, the conflict had been resumed and many young men of Chulmleigh heard the news with some dismay. When they stepped out into the streets, or turned corners, they would have been careful to take a quick look to make sure no press-gang was approaching. On the signing of the Peace Treaty many of the Royal Navy's ships of the line had been hastily laid up (too hastily, some thought) and their crews paid off. Now all haste was being made at Plymouth, Bideford and all other ports to recommission, refit and re-man for war and, because of the dreadful conditions in which sailors had to serve, personnel were hard to come by. The very real threat of a French invasion made many men reluctant to join either the army or the navy and leave their wives and children at the mercy of the enemy, whose behaviour towards defeated opponents was said to have been barbarous in the extreme.

On that sweltering August morning the little town must have presented an animated picture to any traveller arriving from Exeter or Barnstaple as their stage- or mail-coach drew up outside one or other of the six inns: the Barnstaple Inn, the Black Horse, the Globe, the King's Arms, the Lamb and, probably the most important hostelry, the Red Lion, which provided fresh horses for coaches.

For more than four hundred years Chulmleigh had played its part in the great woollen-serge manufacture that had been the county's biggest source of employment, but this trade had so declined by the end of the 18th century as to provide work for only a handful of the inhabitants. The population of the town in 1803 was about 1,400 and of those suitable for the work a great many must have had jobs in the inns themselves or in the business that serviced them. Stable boys, ostlers, grooms, hire-chaise drivers, waiters, cellarmen, cooks and chambermaids worked all day and all night to serve the needs of customers.

Children must have thronged the stifling streets on that summer day, schooling not then being compulsory, all darting here and there among the carriages and carts in the hope of earning a copper for holding a horse or running an errand. On market-days they helped the farmers and stockmen to round up straying cattle or sheep, or to catch escaping piglets and poultry. Some minded fat lambs in pens hastily put up in fields near the auction ring until the time came for them to be sold. On market-day the streets of Chulmleigh, especially those adjacent to the sale ring, must have sounded like an over-crowded abattoir.

Such small towns as Chulmleigh were, at the beginning of the 19th century, largely self-supporting, usually with about eight per cent of the population on Parish Relief while the rest of the workers provided the goods and services the town required. There were, for example, several carpenters, a couple of cabinet-makers, two wheelwrights and coopers, all working with locally grown timber and each employing one or more assistant as well as a couple of apprentices. Such a busy town with its 'passing trade on wheels' would have had as many as three blacksmith's shops where four or more men and boys would be kept busy from daybreak to dark, all the year round. Practically every article in general use was hand-made so that, although wages were low, there were usually jobs to be had, especially in spring, summer and much of the autumn when the farms would have absorbed any spare labour to help get in the harvest.

On Fridays the town must have been extra busy. Carts and pack-horses would have started arriving at dawn, and cattle, sheep, pigs and poultry would have added to the smells and noises echoing through the streets together with the shouts and curses of cowmen and shepherds. As the morning wore on the usual crop of daily vehicles would converge on the town and coachmen, particularly the drivers of mail coaches, who prided themselves on their punctuality, would have added their shouts and the trumpeting posthorn would have swelled the tumult converging on the market place, the Red Lion or one of the other inns. Most of the cottages had back yards and outhouses and many of the occupants would have earned a welcome shilling by hiring them to the farmers.

Looking today at the silent and peaceful little town, long since bypassed by turnpike and railway, it is perhaps difficult to imagine the hurly-burly, the urgent to-ing and fro-ing of the inhabitants that once had been commonplace. Thinking of the stench and the incessant tumult in the streets, now as quiet as the little graveyard beside the church, makes the early years of the 19th century seem remote indeed.

At the foot of the hill upon which Chulmleigh is built flows the sparkling stream called the Little Dart, which winds its way through the trees and round the meadows until, about a mile beyond the town it converges with the River Taw. These rivers indicate the water-level in the area and, in order to supply the town at the top of the hill it was necessary to dig a well as deep as the hill was high. On the evening of Wednesday, August 17th 1803 the rivers were running low, the Little Dart being scarcely more than a trickle and the town pump in the market-place was almost dry when, soon after 9 o'clock in the evening Mr. Ford, the soap-boiler, had just finished work for the day and was scraping out the embers from under his copper boiler. While he was doing so, someone opened a door, creating a draught which rekindled some fragments and blew them up the chimney. They instantly ignited the dry thatch of the roof.

Almost every house in Chulmleigh was, at that time, constructed of timber and cob with a roof of straw thatch which had become tinder-dry in the long drought. In an instant Mr. Ford's roof was burning like a torch and a rising wind was blowing in eddies which seemed to come from every quarter by turns. Fragments of burning straw from the roof were blown away and quickly set fire to other houses further down the street; then, on a sudden change in the wind's direction the fire swept across the road to ignite roofs on the other side. In an instant another gust had swept lighted particles up the hill and Mr. Ford's house was completely surrounded by fire. Such was the dry state of everything with which the flames came into contact that twenty house roofs were ablaze in a matter of minutes nor was there the remotest possibility of quenching the fire as the town's principal water supply was virtually non-existent.

When the alarm was raised, many families would have been in bed and had to escape in only their night attire. Terrified children waited for parents who had returned into their blazing homes in the hope of salvaging their most cherished possessions.

In stables and yards behind dwellings and shops, men struggled to pacify frightened horses or to extricate squealing pigs which, when released into nearby meadows, rushed in headlong flight from the blazing town—many never to be seen again by their owners. Quickly the narrow streets became filled with choking, acrid smoke from the smouldering thatch, while anxious people, bewildered and frighted, jostled one another as they tried to push loaded barrows or to pilot sick and infirm elderly relatives to safety in the fields beyond the houses or into the comparative safety of the parish church. Those who had handcarts loaded them with their furniture and perched small children on the top. Such was the rapidity with which the conflagration spread that these handcarts or barrows caught fire as they were pushed over clumps of burning straw so that the children had to be snatched from their place and carried through the licking flames. Many suffered from burns to the face and head while others, especially the unshod children, were badly burned on the soles of their feet as they ran here and there in search of their parents. The shrieks of the children, the screams of the women and the shouts of the men added to the utter chaos of the horrifying scene as they ran through the dense clouds of choking smoke bumping into each other, none seeming to know what to do or which way to run. In the narrowest streets the flames barred the way and the cottagers had to scramble over back-yard walls to safety. A kind of hysteria gripped the people who, shocked by the awful spectacle of the holocaust that had destroyed their all, were incapable of constructive, co-ordinated action, nor could they attend to anything beyond their immediate need to escape with their lives and save their children. The newspaper, the *Exeter Flying Post*, in its issue dated 25th August published the following account:

> Whereas a dreadful fire broke out by accident in the middle of the town of Chulmleigh, on Wednesday evening, the 17th instant between 9 and 10 o'clock, and from the long drought and extreme scarcity of water contiguous to the dwelling houses and outhouses, all of which were covered with thatch, the fire continued raging in an irresistible fury for between 8 and 9 hours and reduced to ashes 95 dwelling houses. This number being more than two-thirds of the town, the value of which, together with other property consumed was, according to the best estimate at present available, between twenty thousand pounds and thirty thousand pounds.

This report was followed by a plea for subscriptions on behalf of the victims.

A committee was appointed as soon as possible for the purpose of ascertaining the real losses, and other circumstances of the sufferers and of stating to the public the amount of the benefactions received.

Some seven hours after the outbreak started, as the summer dawn began streaking the smoke-laden sky with daylight, the fire had at last burnt itself out but not before a row of cottages had been demolished to make a fire-break. Almost 500 persons had been rendered homeless. It is remarkable that not one person perished in the blaze although several were severely injured, mostly in heads and hands, the latter caused by beating out their own burning clothing. Several may have lost nearly all their hair through burning thatch falling from roofs as they ran by. The phrase used in the newspaper appeal '. . . reduced from a decent subsistence by industry to the utmost distress . . .' draws attention to the practice, common at that time, for employed persons in many crafts and industries to supply their own tools; even farm labourers were expected to have their own hand implements while carpenters, joiners, wheelwrights and similar craftsmen were expected to have a sufficiently comprehensive set of tools as to enable them to tackle any job required of them. In the fire on the night of 17th August many craftsmen's tools were destroyed and their owners therefore prevented from earning their living until replacements could be obtained.

On that dreadful morning following the disaster men must have looked on the blackened ruins of their homes with a feeling of deep despair. There was so much to do and no tools for the work but, as the morning advanced and news of the conflagration reached nearby towns, help began to arrive. The stunned inhabitants, moving about the ruins in a sort of numbed silence, were suddenly cheered by the cries of wagoners, the pistol-shot crack of the carter's whip as convoys of vehicles brought relief from all directions. Laden carts from South Molton, gigs from Bideford, carts from Torrington, wagons from Crediton and North Molton came lumbering into the market-place, each piled with food, blankets, tents, clothing, hams and barrels of beer but, most important of all, they brought tools of every kind, to say nothing of willing hands to help demolish tottering chimneys, or to clear away the rubble and to assist the people of Chulmleigh in the task of rebuilding their town.

The landowners from the surrounding district sent in their agents, or rode in themselves to assess the town's needs for timber, etc., and in the space of a few days, reconstruction was already started.

Meanwhile the parish church had to be used to house many families, especially those with old folk to whom exposure in the fields might have been fatal. Those whose houses had been fortunate enough to escape destruction took in as many as their rooms could accommodate. The prevailing clement weather helped the situation, too, as many of the more robust were content to sleep in improvised tents or in borrowed army ones erected in the fields and the business life of the town was interrupted as little as possible. Only two days after the fire, the Friday market-day was not entirely abandoned and one can imagine the auctioneer standing on a heap of blackened bricks while the hooves of the restless cattle stirred up clouds of sooty dust. The mail coaches, too, would have clattered into Chulmleigh just as usual on their way to Barnstaple, Bideford and Torrington, or on their journeys from those places to Exeter.

The Red Lion Inn, luckily undamaged in the blaze, attended to the needs of passengers and provided the necessary change of horses. The sight that met the gaze of travellers on that morning must have been horrifying as, from their conveyances, they saw weeping women whose children wandered forlornly among the ashes seeking anything that might have escaped destruction, to keep as a talisman for the future. Many of the travellers dipped deep into their purses that day, moved to pity by what they saw and a humane desire to relieve the victims and they would surely have talked about Chulmleigh's plight wherever they went.

Many of the shops and houses in Chulmleigh, especially those owned by the lord of the manor, were, of course, insured against fire so that a high proportion of his rebuilding costs were adequately covered. Few of the smaller landlords or owner-occupiers would have been so protected and fewer still would have thought to insure their goods and chattels against fire or even their tools of trade and it was for these people that the Subscription List was started. The committee set up to administer the fund calculated, after much deliberation and careful questioning of the inhabitants, that the uninsured losses amounted to about £11,000, a large part of which was accounted for as shopkeepers' stock which had been destroyed, while furnishings and household goods amounted to a smaller sum as a great deal of the furniture belonging to the poorer people would have been improvised or home-made.

By November of 1803 a published statement showed that the Committee had paid to thirty-three families the entire amount of their claims, none of which had exceeded £5. In fact the total sum paid to these claimants amounted to only £81, an average of less than £2 10s per person. In spite of the efforts of the committee, however, the first rush of subscriptions from the public was not maintained and it took two months to raise the first £1,000 and not until 10th November was that sum doubled.

The final statement was issued by the committee fifteen months after the fire, on 19th November 1804, by which time the subscriptions received amounted to £3,645 12s 7½d. As the uninsured loss was reckoned to have been about £11,000 this response must have disappointed the people of Chulmleigh and would have been a considerable financial setback. In the circumstances the best the committee could do was to see that the greatest burden of loss fell upon those best able to bear it, as the following announcement in the *Exeter Flying Post* makes clear:

> The Committee who undertook the enquiry into the losses of the sufferers by this fire, and the distribution of the donations for their relief, hereby close their accounts by a general statement of their proceedings. They are happy to find that the amount of the subscriptions, amidst the heavy demands of the times (the war with France, etc.) does so much honour to the public character and affords such essential benefit to the sufferers. In the distribution of the subscriptions the Committee have generally allowed to the sufferers of the lower classes whose losses did not exceed £20, the whole of their losses and the allotment to the sufferers of the higher classes have been regulated by the amount of their respective losses on a fair estimate considered with reference to remaining resources, families and other circumstances. It has been the great object of the Committee, in the discharge of their trust, to ascertain as far as it was possible, the real merits of every application and to render impartial justice to the several sufferers by an equitable distribution of the public munificence.

The statement bears the signature of the chairman, and is dated Chulmleigh, 19th November 1804. Thus it is evident that the ‘higher classes’ had to share the loss of around £7,000, a salutary lesson to insure against the risk of fire.

Considering that such a large proportion of the town was destroyed, including practically all the shops, the fact that the committee was prepared to close the appeal fund after only fifteen months and when less than £4,000 had been subscribed, would seem to suggest that the people of Chulmleigh had, in that short time, reconstructed not only the devastated buildings but their businesses and social lives as well: surely a quite remarkable feat of determination by a resolute population and they surely deserved better than the fate that awaited them in the not-too-distant future. The first irreparable blow to strike Chulmleigh was to occur in 1823 when a new turnpike road was constructed which bypassed the town as it followed the valley of the Little Dart so that the lumbering coaches and heavily-laden carters' wagons no longer had to climb the steep gradient into the town, so that the all important 'passing trade' was lost for ever. An even worse blow was to fall on Chulmleigh's commerce when, in 1854, the railway came to North Devon, for it, too, followed the easier valley route and new, conveniently sited cattle markets were set up at South Molton Station and down at Eggesford. The last big cattle market was held in the town soon after and, turning its back upon a world which must have looked ominously unfriendly, a future empty and bleak, settled down to nurse itself through the coming century with a quiet fortitude that has left the little town unspoiled and determinedly tranquil.

When the town was rebuilt after the fire, in 1803, thatched roofs were replaced by slates for safety although here and there are houses still covered with picturesque reeds or straw and it is not difficult, with these reminders at hand, to guess at the appearance of the old town before the fire.

2

CHUDLEIGH

The construction of a modern dual-carriage bypass and, later, the M5 motorway which strides across the Exe valley to carry its wide thoroughfare westward have, together, brought to the small market town of Chudleigh that peaceful quietude only to be enjoyed by those with time 'to stand and stare'. Long before either of these roads had come into existence, however, Chudleigh's main street was part of what was then known as the Great Plymouth Road which, during the tumultuous war years at the beginning of the 19th century, was an exceedingly important and busy road fitting perfectly the conditions so aptly described by George Eliot in *Felix Holt*, *the Radical*, 'Five-and-thirty years ago the glory had not yet departed from the old Coach Roads. The great roadside Inns were still brilliant with well-polished tankards, the smiling glances of pretty barmaids and the repartees of the jocose ostlers. The Mail still announced itself by the merry notes of the horn. The hedgecutter or the rick-thatcher might still know the exact hour by the unfailing, yet otherwise meteoric apparition of the Pea-green "Tally-ho" or the yellow "Independent". . . .'

The town of Chudleigh lies, as it were, on the hem of Haldon Hill's long skirt, on pleasantly undulating ground surrounded by picturesque countryside. It is nine miles from Devon's county capital, Exeter, a convenient distance for the mail coaches to make the first change of horses en route to Plymouth and, of course, the final change for coaches travelling in the opposite direction.

Like so many of Devon's small market towns, Chudleigh had, for several centuries, depended for its livelihood on the woollen industry which, however, by the year 1800 had shrunk almost to nothing although a few weavers still operated their looms in outhouses behind cottages. Two large mills in the town stood empty, their roofs and walls decaying, their once busy, noisy machinery long since removed. The people of the town had found other employment in the nearby lime kilns and stone quarries and in supplying all the varieties of services their fellow townsfolk needed. Not the least important employers of labour were the two inns on the Plymouth Road, The King's Arms and the Plymouth Inn, the former probably having stabling for as many as fifty horses as well as covered coach-houses to accommodate many private vehicles.

Besides its parish church, the town boasted a school and a fine Play Park where the children could romp in safety, though few over the age of eight years would have had time for play, instead spending the daylight hours doing whatever work they could get. There must have been plenty for them, too, with the constant stream of travellers on horseback or in wheeled conveyances who would send the children running to fetch grooms or ostlers.

In 1807 preparations were going forward at Plymouth for transporting an army into the Iberian Peninsula and the great road through Chudleigh must have been continually congested from 3 in the morning, when the London Mail arrived from Exeter on its way to the Naval Dockyard at Plymouth. The main road through Chudleigh must also have resounded to the tramp of marching soldiers from the barracks at Exeter, together with their cartloads of equipment and wagonloads of food and ammunition all intermixed with the usual private and commercial traffic.

The spring, that year, had been unusually dry, almost no rain having fallen throughout the month of May. A little before noon on Friday, 22nd May, a fire broke out at the premises of a baker in what was then called Culver Street. Some kindling had caught light and sparks had been blown by a stiff breeze up on to the thatched roof. In an instant it was blazing furiously while the wind whipped up strands of burning straw and carried it to the roofs of adjoining cottages and across back yards to set light to thatched outbuildings. In a matter of a few minutes three streets of dwellings were well alight.

Among the first casualties to fall victim to the blaze was Chudleigh's only fire-engine, a contraption built in the form of a large wooden tank on wheels. In the centre of the tank a brass pump was fixed; a clumsy affair requiring about a dozen strong men to operate the heavy shafts that were the pump-handles. On that day it was not brought into use because, after the long spell of dry weather, the town's water supply was practically non-existent.

Chudleigh got its main domestic water from the hills by way of a leat and distributed it about the town through earthenware pipes and culverts or, as in most of the smaller streets, simply in open ditches which ran along before the front doors of the dwellings. When the fire started there could have been scarcely more than a trickle of water in the greater part of the town. Some houses, especially the larger ones occupied by the wealthier residents, would have had their own wells but it would have taken far too long to get water from these, for any practical fire-fighting attempt to be made, even had the wells been full, which, in the circumstances, was highly unlikely.

The fire spread so rapidly in the freshening wind that people were frightened almost out of their wits. One moment they were staring at billowing smoke rising from the far end of the street and the next minute flames were crackling all round them, their own homes being among those with blazing roofs. They hardly knew which way to turn or what to do, so many things demanding their immediate attention. To begin with it was urgent to find the children, especially babies, and many of the cottages had outhouses containing livestock; a horse, perhaps, or pigs and poultry. Behind some of the dwellings were small orchards which provided apples for the town's cider and the grass between the apple trees provided grazing for a milking cow. If the people were terrified of fire the animals were even more frightened and it must have required much care to get the creatures safely from burning stables or sties out into the surrounding fields. When the fire started those living near the outbreak took their animals to the orchards of friends whose property they thought to be safe, only to have to remove them again within a few minutes.

In a book written by an inhabitant of Chudleigh at the time of the fire, one Mary Jones, there is a story of an old man who had died the day before and, when the house in which he lay in his coffin caught light, it was moved by neighbours out into the street. His son, who lived at a little distance in Mill Street, was told and at no small risk to himself, rushed through the flames and found articles piled about the coffin already alight. With the help of two friends he carried the coffin to the churchyard, managed to find the vicar, and an exceedingly hurried interment took place, the grave having been dug the night before.

Babies and old folk were moved from one house to another as the flames followed them from one side of the town to the other and, in the narrower side-streets, escaping people had their clothing set alight by the licking flames. At one time it had been thought safe to load precious belongings into barrows and handcarts and to take them to the Market House in the square, a wooden building not more than three or four years old, and one of the few buildings in Chudleigh at that time which had a slated roof. This feature seems to have given the inhabitants the idea that it would not take fire. In spite of its being of timber construction they might have had better fortune had they taken the elementary precaution of making quite certain that the goods they deposited therein were not, themselves, smouldering. Draughts soon fanned the smoking articles into flame and soon the Market House was burning furiously while women ran hither and thither wringing their hands and bemoaning their losses. The scene must have been chaotic in the extreme.

In the 1801 Census, the first ever taken in Britain, Chudleigh was shown to have had a population of slightly in excess of one thousand seven hundred souls and all but a few who worked away from the town, must have been milling about its streets wherever it was still possible to pass without being roasted alive. The main road to Plymouth was soon blocked and coaches on their way to Exeter had to be diverted through the recreation ground, known as the 'Play Park', and out into the narrow lane beyond in order to bypass the burning ruins. This must have been quite a feat on the part of coachmen and guard and it is not difficult to imagine them trying to calm the terrified horses. The guard would have had to throw a blanket over the leaders' heads so they could not see the flames they had to approach before turning from the road. The coachman, too, must have needed all his skill to hold in the wheelers as they moved down the hill nearer and nearer to the blazing houses, heads tossing, nostrils flaring in terror, tails lashing and hooves dancing a fandango of fear while the coachman leaned over them, both hands holding the reins now as he coaxed the sweating creatures onward a few more yards and a few more until at last the guard, stumbling along backwards as he held both jerking heads, was turning the leaders away through the Park gates and to the fresh, clear air. Then the sudden silence as the rumbling, crunching wheels of the coach ran on to the grass. The coachman had to nurse his team on for an extra nine miles to Exeter because the King's Arms, where he normally changed horses, was ablaze with its outhouses and stables, too.

At about 2 o'clock, that is, some two hours after the fire began, Chudleigh was rocked by a tremendous explosion which sent burning debris high into the air and scattered if for a considerable distance and, no doubt, must have brought down the chimneys of many of the nearby ruins. A man employed at the quarry or lime kilns had a barrel of gunpowder stored in an outbuilding of his cottage in the very centre of the town. He had either forgotten it or, which is the more likely explanation, had been too frightened to attempt to move it through the burning steets to some place of safety.

The panic this additional disaster caused may well be imagined for the explosion occurred at a point about the very centre of the town where the houses were of the smallest size, many doubtless consisting of only one room on the ground floor and one above. The poor, in those days, expected neither space nor privacy yet they seem to have possessed a capacity for making the best of their lot while landlords probably felt that, if poor people wished to have large families, it was no business of anyone but themselves. In such households (or should they be called hovels?) there would have been little furniture, and most of what was there would have been home-made. Few of the very poor had bedsteads but slept on straw-filled sacks laid on the bare floor. To such people the fire would have caused little financial loss but however sparsely furnished their dwelling might have been, it was home, a refuge, a protection against more than inclement weather, a place, perhaps, to hide their poverty.

'Upwards of 1,000 persons of all ranks, men, women and infants; the aged, the sick and the infirm all at once deprived of a home' said the *Exeter Flying Post* in its issue of 28th May, 1807. These victims of the fire found themselves entirely at the mercy of the few whose houses survived and of friendly farmers in the surrounding countryside, many of whom hastily cleared barns or other outhouses to provide temporary homes.

As the fire spread, the inhabitants were driven further and further away from the town centre to the outskirts, so that when the explosion occurred none was near enough to be in immediate danger either from the blast or from crashing chimney-stacks and collapsing walls.

It is interesting to observe that Mary Jones, in her account of the explosion, tells us that the cause was 'a barrel of gunpowder' but in an account in the *Exeter Flying Post* the following appears: 'The town of Chudleigh, nine miles from hence, has been consumed by fire. The Clifford Arms and The King's Arms with the three other Inns are all in ashes. During the conflagration several explosions of gunpowder took place as a quantity was kept in the town for blowing up rocks.' A week later another account, in the same newspaper, reads: 'On Friday, 22nd May, about 12 o'clock, a fire broke out in a bakehouse in the town of Chudleigh which raged with increasing fury, and, at length, communicated to a house containing two barrels of gunpowder.' Perhaps the tendency for the press to incline towards exaggeration had already begun. 'Several explosions', however, may have seemed to be going a little too far, but the editor was evidently reluctant to retreat to anything quite so unsensational as a single one when, after all, his job was to stimulate his reader's interest.

The uncoordinated efforts of the dazed inhabitants did nothing to confine the spread of the fire and even further damage must have ensued had not the direction of the wind shifted while, at the same time, a number of men from nearby villages and farms, arrived on the scene. Not being personally involved in the loss of their homes or possessions they were able to take a more detached view of the frightful situation and realized at once that it was necessary to demolish some houses in the path of the flames to make a fire-break, with the many hands available, a number of houses were quickly stripped of their contents and pulled down. By four o'clock, having raged for more than four hours, the fire was at last brought under control and the flames were beaten out.

The sight the poor townsfolk faced as evening shadows lengthened was horrifying. The entire centre of Chudleigh was nothing but gaunt chimney-stacks and blackened walls, still hot and stinking and in imminent danger of collapse, so that to walk in the narrow streets was to court death. Quantities of the old, rotting thatch still blocked the thoroughfares, hot and smoking and giving off a stench like some fetid cesspit in the heat of high summer. What remained of the buildings would, it was all too plain to see, need to be demolished as soon as possible and the blackened, crumbling rubble of the old cob dwellings cleared away.

That there was plenty of work for all who cared to do it was in no doubt but, they must have asked themselves, on what would they live while the work was in progress, and, furthermore, from where would the necessary materials come for rebuilding the town? They would have been aware, no doubt, that the largest landlords would employ them to rebuild the houses they owned. Although what proportion of the devastated property these owned was not clear, the landowners and gentry, already alerted, moved into action.

On Saturday, the day following the conflagration, a meeting was called at the vicarage and was attended by as many of Chudleigh's inhabitants as could get in. No doubt, in such clement weather, many stood in the garden to hear the proceedings. Lord Clifford was proposed as chairman of a committee to be elected to solicit funds for the relief of those who had lost property which had not been insured, and to make what arrangements were possible for the immediate care of the homeless.

The committee, formed under the Baron's chairmanship, had among its many members Sir Lawrence Palk, M.P., the Rev. Gilbert Burrington, Vicar of Chudleigh, and some eight or nine other ordained priests, many of whom were absentee incumbents of parishes scattered far and wide and who lived on the stipend while paying a pittance to a curate to run the parishes.

Among immediate matters attended to at this first meeting was to record a vote of thanks to all the neighbouring towns and villages from whence food and clothing were pouring into

Chudleigh. Cartloads of bread, beer, dairy produce and blankets arrived almost hourly to be distributed by the hard-pressed vicar in the Play Park to those in need. The commanding officer of the military barracks at Exeter sent tents which were quickly erected in the surrounding fields and the business of demolishing the most dangerous ruins was begun.

Donations from the nobility and gentry of the surrounding district began to come in and, at that first meeting, a vote of thanks was passed to a Mr. Collins of the Devon Bank and to Mr. Nation, of the City Bank, Exeter, for their offer to advance whatever money was immediately needed for the relief of the distressed sufferers. Among the places to which money could be sent for the fund was Hoare's, Bankers of 17 Beer Street, London, Alexander Davison's at 11 Pall Mall and Hobhouse & Co. of Bath as well as most banking houses in the counties of Devon and Cornwall.

The news of the disaster reached Exeter while Chudleigh was still burning, and the mayor lost no time in issuing an urgent request to all parishes in and around the city to institute a house-to-house collection immediately and the response was quite remarkable, even the poorest parishes collecting comparatively large sums. On the Tuesday following the fire the Rt. Worshipful Samuel White, Mayor of Exeter, announced that donations received from the parishes included £67 from the 'Precincts of the Cathedral Close' alone, while St. David's had collected £73, St. Mary Major £54, St. Paul's, Bradninch £67, St. Sidwell's £119, Holy Trinity £81 and there were many more. Subscriptions were also collected from many sympathetic private citizens at banks and the newspaper offices.

The days following the fire were difficult for the inhabitants of Chudleigh. As they endeavoured to get on with preparations for rebuilding the town, an influx of sightseers stood around and stared dumbfounded at the desolation, constantly, one imagines, getting in the way of sweating men trundling away barrow-loads of debris from demolished walls and neither the acrid smell that lingered in the ruined streets nor the choking dust of the demolition deterred these visitors from satisfying their morbid curiosity. Someone, however, had the bright idea of setting up collection boxes at both ends of the main street and there seems no doubt that the sightseers gave generously, but in the published lists of subscribers there seems to be no record of the money collected in this way so it can hardly be unreasonable to suppose that there was a sort of general share-out among the people of Chudleigh, at any rate the poorer classes, by those who initiated the scheme. The various regiments of militia stationed in the vicinity also made collections, both among officers and other ranks, which added substantial sums to the fund. The management, staff and actors of the Exeter Theatre gave an evening performance without wages so that the total of box office takings could be contributed. The proceeds amounted to £32 11s 0d.

The earliest published estimate of the cost of the damage done in the great Chudleigh fire was £70,000 but this was later modified to £60,000, the greater part of which was covered by fire insurance. It is known that Lord Clifford was one of the landlords who possibly owned most of the houses, but there seems to be no record as to who owned the rest of the property in the town. It is possible that Sir Lawrence Palk may have been another landlord whose property had been ruined but, be that as it may, it is on record that, very soon after the disaster (later in that same year, 1807), together with others he founded the West of England Fire Insurance Company, which commenced business with a capital of £600,000. It was the first fire insurance Company in Exeter to purchase its own fire-engine, a machine which was to gain fame and the affection of the people of Exeter who referred to it, for the next eighty years of its working life, as 'The Little West'. Its brigade was formed from men who were keen, efficient and inordinately proud of their engine and its reputation for always being first to arrive on the scene of any fire. The old machine is still preserved. Such fire-engines as The Little West were manhandled to local outbreaks and only drawn by one or more horses if the distance it had to go exceeded a mile or two.

The subscriptions to the relief fund came in from all nearby towns. The Mayor of Plymouth inaugurated a collection as did other civic leaders and by Christmas the sum of £15,000 had already been received and the final total was to reach £21,000.

In her *History of Chudleigh* Mary Jones observes that the committee, under the chairmanship of Lord Clifford, had some difficulty in arriving at the true value of losses sustained by the many claimants and the task of allocating money was made no easier by a state of strife and disunity among the committee members themselves. News of the large amount of cash subscribed encouraged the more crafty among the inhabitants to be in no hurry to reveal their estimate of personal losses, or to reassess those already submitted, but some spoilt their chances by inflating their valuation of loss beyond the bounds of probability. Such people were, evidently, not numerous and it was stated in the treasurer's accounts that 409 persons were paid compensation amounting to sums from as little as a few shillings to many hundreds of pounds. The two largest sums paid were £1,194 3s, to a maltster named Wright, while Richard Rose, landlord of the premier inn, was paid £1,207, with which he rebuilt his hotel on a site near the original.

There was no lack of enterprise in Chudleigh once the initial shock of the disaster had been absorbed as the following advertisement in the *Flying Post* just one week after the event shows:

> John Weston of the Clifford Arms respectfully informs the nobility and gentry and public in general that he has entered upon that roomy and commodious house which stands in the west end of the said town of Chudleigh, and is now ready for the immediate reception of company. He takes this opportunity of returning his most grateful thanks for their patronage and favours which they have been pleased to honour him with and humbly solicits the continuance of the same which he will make it his duty not to forefeit. Dated Chudleigh, 29th May, 1807.

The Clifford Arms was, at the time of the fire, one of the smaller alehouses situated in a small street off the Great Plymouth Road and unlikely to have had much of the coaching trade. It was completely destroyed on 22nd May.

Weston's example was not lost on Richard Rose, landlord of the King's Arms, who, shortly after, found himself temporary accommodation, as an advertisement in the *Exeter Flying Post* of 9th July bears witness. In it he told his customers he was ready to give them his usual services, attend to their comforts and change their horses; thus valuable business was kept in Chudleigh as well as employment for many of the inhabitants during the rebuilding period.

At this point it may not be inappropriate to consider in greater detail the state of disunity reported to have occurred among the members of the committee and try to discover any clue as to its cause. Some unknown members evidently came to the conclusion that an Act of Parliament would be an advantage giving the committee powers to realign streets and settle all disputes over pieces of land, etc., which could otherwise delay rebuilding.

In view of the experiences of other Devon towns, in which large areas had been destroyed by fire, and in which few difficulties seem to have arisen during rebuilding, it would seem that somebody was taking the proverbial steam-hammer to crack a rather small nut. While to suggest that such a Parliamentary Bill would be to the benefit of the larger landlords might be uncharitable, it cannot be said to be altogether unreasonable. Such an Act would have required some time to draft and even longer to steer through Parliament where, no doubt, Sir Lawrence Palk, a member of the Lower House, would have been invaluable. Any difficulties were brushed aside, however, and the Bill received Royal Assent on 1st June 1808 and was entitled 'An Act for the Better and More Easy Building of the Town of Chudleigh in the County of Devon and for Determining the Differences Touching Houses and Buildings Burnt Down or Demolished by the Late Dreadful Fire There and Preventing Future Damage by Fire'. The Act lists the entire committee as commissioners and carefully states that seven members would be sufficient to make absolute decisions, against which there would be virtually no appeal. This seems rather remarkable. Were such drastic measures really necessary among such 'honourable' gentlemen?

If we look at the few statistics available concerning Chudleigh we discover that, before the fire there were 300 dwellings and that some 180 of these were rendered uninhabitable. After rebuilding was completed, by the year 1811 the number had grown to 370. In spite of this large increase Mary Jones tells her readers that some of the pre-fire streets, such as Mill Lane, were very narrow indeed but that after the fire these were widened, actually having a pavement, or footpath, running along at least one side.

If the streets of Chudleigh were thus widened and seventy more houses were erected where, it may be asked, did they find the extra space? In all the existing accounts of the conflagration there are frequent references to the yards behind the houses and suggestions that the outbuildings therein were of more importance than the dwellings themselves. What is more often stressed is that these outhouses were constructed, as often as not, of stone or brick whereas the cottages were timber and cob with stone chimneys. In addition to these outbuildings the accounts also make references to small walled orchards behind many houses. Of course the word 'small' is comparative and to be worth-while as an orchard these plots must have been large enough for three houses. Some, if not most of these orchards may have been attached to properties with owner-occupiers, as Mary Jones tells us was the case with the post office. The orchards provided cider, the staple drink of the poorer townsfolk, and to uproot the apple trees to build more houses would surely have upset the traditionalists.

The Act's main purpose seems to have been to give compulsory purchase powers and to enable redevelopment of the town by the principal landlords.

The point seems to have been that, with the powers which the Act conferred on the committee, it could not only insist on an owner parting with a piece of his land but could also decide the price he should be paid for it. It is hardly surprising, therefore, that among the 'honourable' gentlemen of the committee, many of whom were ordained priests, one or two might have experienced the occasional pricking of a conscience and may even have been bold enough to voice their misgivings in committee. If any abuse of the powers vested in these gentlemen took place, there was nevertheless no public outcry. Mary Jones suggests that it was a pity that the people of Chudleigh did not see fit to erect some monument to commemorate the great kindness and charity they received from the public as a result of the fire but it is possible that at least some of them may have harboured a sense of grievance rather than gratitude. Those, for example, whose orchard had been reduced to a cabbage patch, or whose pigsty had been demolished because it stood in the way of somebody else's 'improvement'. The poorer people were by no means incapable of taking advantage of the situation and the committee members had great difficulty in checking claims by the more cunning elements. A report addressed to the chairman, Lord Clifford, from the treasurer read: 'In adjusting claims so multifarious and complicated it cannot but be supposed but that some impositions may have been practised with some success.'

In the second and revised edition of her *History of Chudleigh*, Mary Jones remarked 'On an inspection of the list we must feel surprised at the immense sums awarded to particular individuals and also at the names of many wealthy persons who participated in public charity. It cannot but be regretted that the Committee that had so much power and money at its command had not exercised it in securing for the town buildings of architectural strength and uniformity.' It is regrettable that she considered it unnecessary, or unwise, to mention names in this case and, what might have been even more interesting, the status or professions of the persons to whom she referred.

There is another matter upon which there seems to be no available information; it is that of interest on the money raised. The published list of subscriptions to the relief fund had passed £15,000 by Christmas, 1807. Certain sums had, as already stated, been paid in compensation but it seems reasonable to suppose that there may well have been as much as £10,000 left in hand at that time, and held to defray the cost of rebuilding uninsured property. According to Mary Jones the reconstruction was completed by the end of 1809 but the final payments seem not to have been made until 1811 so that there could have been accumulated interest on the money to an amount of up to £1,000 or more because the final total of subscriptions published was £21,000. The disunity among the members of the committee may have resulted from someone asking awkward questions concerning such interest. Bankers, after all, like to make money earn and if such a sum as was here involved was permitted to lie idle it was extremely bad business. On the other hand, if the money had been invested in some such safe commodity as Government three-per-cents, where did the resulting interest payment go?

Perhaps there is a shadow of anticipated doubts in the committee's resolution, passed on June 1st, only nine days after the fire and which read as follows: 'It is resolved that no person who claims any benefit by the late Fire, at Chudleigh, be a member of the Committee. The Committee deem it highly necessary to do away with any wrong impression respecting the tenure of the property destroyed. It is further resolved that the public be informed that the houses destroyed were Freeholds and that no individual constituting the Committee has any interest whatsoever in the subscriptions and that the foregoing resolutions be printed.'

The wording of these resolutions surely suggests that either some unpleasant innuendos were already circulating, which would hardly be likely so soon after the terrifying event, or someone with an eye on future transactions thought it expedient to forestall any doubts that might arise.

The final meeting of the committee was reported in the *Exeter Flying Post* as having been held at the Vicarage, Chudleigh, on 7th February 1811, when the accounts were closed and the committee dissolved with all the appropriate votes of thanks. An echo of the great fire occurred in the same newspaper, on 16th June 1814, in the form of an announcement concerning the Act of Parliament. It read: 'Notice is hereby given that the Commissioners named in an Act of Parliament entitled 'An Act for the Better and More Easy, etc. . . .' it is intended to meet in the Vestry room in the said town of Chudleigh on Thursday, 30th June by 11 o'clock in the forenoon for the purpose of carrying the said Act into execution. By Order of the Commissioners'. The announcement is signed George Dolling, Clerk to the Commissioners.

The terms of the Act are thus shown to have been implemented and the matter at an end but no details of this meeting seem to have been reported in the press. Towards the end of that year an advertisement appeared in the same newspaper offering for sale, by private contract, 'a commodious Residence with an acre-and-a-half of pleasure grounds and a walled garden, at Chudleigh'. The property was stated to have been erected but a few years since and below the advertisement appeared the signature of the same Mr. George Dolling, the Attorney of Chudleigh. Perhaps the owner had suddenly found himself short of cash or was it one of the houses used temporarily as an inn and no longer wanted?

Happily none of these speculations can ever be resolved because they were not publicly asked at the time and all the people who played their parts in the drama have long since been laid to rest, some, no doubt, in a quiet spot in Chudleigh churchyard. We who come to disturb their secrets can only be grateful to them for leaving behind a legacy of intriguing mystery to enthral the inquisitive. The curious can never have more than a hint to go on and can hope to get no nearer the solution than a guess, a guess that can hurt nobody because the answers in themselves no longer matter.

Mary Jones herself left a riddle for us to attempt to solve. In her book she talked of having seen six engravings by George Hollis from drawings by Henry LeCourt made previous to the fire in 1807. 'They are beautifully done,' she wrote, 'and in the possession of C. Langley, Esq., and others.' Where are they now, still cherished or perished? Who knows? Someone may still have the plates and the original drawings awaiting rediscovery.

H. De Cort. del. 1795. G. Hollis sc. 1829

VIEW IN CHUDLEIGH,

DEVONSHIRE.

BRADNINCH CHURCH.

Drawn from Nature and on Stone by W. Spreat.
Printed by C. Hullmandel.

3

BRADNINCH

The importance of the First Reform Bill, in terms of political and social progress, cannot be denied but in the year 1832, when it became law, its significance would appear to have been somewhat exaggerated. The purpose of the Act was to reform Parliamentary franchise and it encompassed a sweeping redistribution of seats by doing away with what were known as the 'Rotten Boroughs' and in giving the vote to borough householders paying a property rent of £10 or more per annum, and in the counties, or rural areas, to those occupying properties with a minimum rent of £50.

This new definition of eligibility to vote brought to the polling booths many who had, hitherto, been denied the right but it must not be overlooked that the great majority of the working class paid far less than £10 a year for the dwelling they occupied and a cottage, comprising one room downstairs and one up (the upper room reached by a ladder through a hole in the ceiling of the lower room), frequently housed two families and cost only a shilling a week to rent. Yet the rejoicing at the passing of the Reform Bill seems to have been universal and quite out of proportion to the social advancement actually achieved by it.

Towns and villages organized celebrations that were as bizarre as they were extravagant. An interesting example of this was the great feast that was given just across the county border at Taunton for an estimated 5,000 persons who, since at the time the weather was clement, probably were entertained out of doors. The provisions bought in for this feast were as follows: 7 fat oxen, 625 quartern loaves, 40 bags of potatoes, 500 lbs of salt, 1,500 lbs of flour, 750 lbs of raisins and 640 lbs of suet – these last three items to make 400 8-lb puddings. Nor was food the only element in the celebrations. Also included were 1,875 gallons of beer and 838 gallons of ale together with 2,000 tobacco pipes (clay pipes, no doubt made locally) and 1,000 oz. of tobacco. The poor folk of Taunton may not have got the vote but, on that day at least, they got full bellies. Such celebrations were not, of course, confined to Somerset but were repeated throughout the length and breadth of the country. In Devon, according to a report in the *Exeter Flying Post*, the village of Budleigh was the scene '. . . of lively animation on Monday sen-night (16th May 1832) occasioned by the celebration of the passing of the Reform Bill. Gymnastic amusements, dancing, etc., were continued with spirit to a late hour.' No mention here of feasting or drinking but the imagination is stimulated as to the nature of 'gymnastic amusements.'

Reform Bill year was not, however, without its troubles for the people of Devon, for even as the celebrations were going on an outbreak of cholera was growing to epidemic proportions and sweeping through the villages and small towns and carrying with it an alarmingly high death toll. Against the spread of the disease, and how to keep it at bay, the best the editor of the *Exeter Flying Post* could offer his readers was advice to those not yet afflicted. 'We regret to state that the victims are mainly of the poorer classes' (evidently he would have preferred the well-to-do to have been the victims). He exhorts the sufferers to '. . . be of moral courage, to preserve a cheerful disposition and, as far as possible, equanimity of mind and to pursue their usual avocations and to live on food as concentrated as possible and as much as they can, and not to suffer meals bearing the character of meagreness but, above all, to observe the rules of sobriety.' Such advice might strike us today as impertinent and patronizing but, no doubt, it did no harm even if it did no good since those to whom it was addressed were unlikely to have been literate and would certainly not have wasted their precious money on newspapers. The number of cases of cholera reported in the two weeks to 2nd August, in Devon, amounted to 64, of whom 23 had died.

On the very day in 1832 when this outbreak of cholera in Devon was reported, that is on 18th July, the small market town of Bradninch suffered another sort of tragedy. The weather had been dry and warm for several weeks and at about ten o'clock in the morning a fire broke out at a bakehouse whose proprietor, a man named Richard Haydon, was also the town's constable, a prominent member of the community. By some mischance the bakehouse chimney caught fire and emitted a shower of sparks on to the thatched roof which instantly burst into flame. The fresh breeze soon spread burning strands of straw among the closely packed surrounding houses and their outbuildings almost before the baker was aware that his own roof was alight. It seems somewhat surprising that, as so many fires started in bakeries, some special precautions were not observed, such as having handy a barrel of water and a bucket or two. Yet there is no evidence that bakers took any special account of their vulnerability but, in common with other townsfolk, seemed to look upon the hazard of fire as a natural risk about which they could do no more than take out insurance. It appears to have been a kind of unwritten rule to abandon a house that caught fire and only to try and salvage the most valuable articles in it.

Bradninch had no fire-engine of its own at that date, so the constable immediately despatched a messenger to Cullompton for help but it took more than an hour to get assistance and, when the fire-engine at last reached the town, it was to discover that so meagre was the supply of water available that it was useless. Firemen and inhabitants could only look on at the awful spectacle.

Everybody was in the streets, or helping to get out horses from stables behind the houses, others vainly trying to round up pigs and poultry or help neighbours remove the old, the sick and the babies to the safety of the fields. The fire, meanwhile, raged and spread in every direction and groups of men with hooks on chains began tearing down the thatch on cottages before the fire reached them in the hope of stemming its progress.

By about 3 o'clock in the afternoon the fire had thus been contained and was shortly after beaten out but not before 48 houses, the Guild-hall, the gaol and the Baptist chapel had been reduced to a heap of smouldering ruins.

There was one building in Bradninch of which the inhabitants were particularly proud and this had been anxiously watched while the fire raged. It was the Spread Eagle Inn, a splendid and unusually well-appointed hostelry which brought a considerable amount of business to the town. The inn stood a short distance down the hill, in the main street, from the church gate and its proprietor took pride in the quality of its service to the traveller, and everything within doors was bright and cheerful with many a 'polished tankard and the smiles of pretty barmaids'.

Outside, the Spread Eagle had a unique feature, certain to attract the notice and comments of the visitor. From its main entrance a canopy or *porte-cochère* formed a covered way to the very edge of the road and was supported on elegant pillars. Thus, on a wet day the traveller could step from the coach and walk into the inn without getting wet. The feature of this canopy, however, was that it carried a carefully tended flower garden on its roof, its greenery and blossoms drooping gracefully over the sides and spilling the fragrance of pinks and cloves, honeysuckle and mignonette on the warm summer air. Happily this building escaped the fire and the travellers reaching the inn paused on its forecourt to stare in horror at the blackened, charred desolation of the gutted houses which constituted about one-third of Bradninch.

If Bradninch had not already staged its own particular Reform Bill celebrations, the fire would have spoiled the mood for any sort of merry-making because, unlike the earlier fires at Chulmleigh and Chudleigh, where no human life was lost, this conflagration was attended with two deaths from burns.

Considering the awful pall of heavy, black smoke that must have filled every space, every street and courtyard, it is hardly surprising that there was some confusion concerning the manner in which one man died, or that the newspaper reports differed in some important details.

The Guild-hall and adjacent gaol had been burning for some time before Jacobs, the gaoler, remembered that he had a prisoner in the cells and getting the poor fellow out was no easy matter. However, with the help of several others, the gaoler managed to rescue his charge, a one-armed man who, after losing his limb while working at the local paper-mill, had probably found it hard to earn an honest living with only one hand and was inclined to help himself with the other to what was not his.

The first fatal incident occurred soon after midday while the occupant of the gaol was being rescued. In one version of the story it was said that a man named Ingersent, a millwright by trade, was aged and ill in bed in a cottage adjoining the Baptist chapel.

The flaming thatch had slid down from the roof and completely blocked the cottage doorway. A passer-by named William Rabjohn, and Miller, the octogenarian sexton, tried to pull and kick away at the thatch. Having succeeded in clearing the doorway, the two men entered, reached the old man and dragged him from his bed but the staircase caught light and, in trying to get him down to the street, they let him slip and were unable to find him in the dense smoke. The two rescuers only just escaped themselves, both being badly burnt. Rabjohn recovered but the old sexton died a few days later from burns and shock.

A report of this event appeared in the *Exeter Flying Post* as follows: 'A millwright named Ingersent, in attempting to save his property, lost his life and four persons who have received injuries were conveyed to the Devon and Exeter Hospital'. No suggestion here that the millwright was either old or bedridden. This brief account was published as long as eight days after the event. Perhaps the brevity is more surprising than the lack of accurate information in view of the short distance, only 8 miles, Bradninch is from Exeter, where the *Exeter Flying Post* was published.

There was, however, another Exeter-based newspaper which carried a report of the fire and consequent tragedy: *Woolmer's Exeter and Plymouth Gazette* in the issue dated 21st July, just three days after the event, reads thus: 'A Millwright named Ingersol, in attempting to save part of his property, fell victim to the flames. He was just in the act of entering the door of his house when the whole of the blazing roof fell on him and he was literally buried in fire. The unfortunate man was extricated with great difficulty, the whole of his clothes having been consumed and his body burned in the most dreadful manner. He lingered on in this most heart-rending condition for about three hours when death put a period to his misery.' There is no suggestion in this account that the writer is in the least doubt as to the accuracy of what he says: it is so definite as to read like an eye-witness's testimony and, when referring to the injured people who were conveyed to hospital in Exeter, the report can tell the reader that, 'we are happy to state that none of them is seriously injured.'

The true story of the unfortunate Ingersent, or Ingersol, whichever was his name, and the precise manner in which he met his death, may well have been something of a mixture of the differing reports. There is a ring of certainty about the last version which seems to rule out the idea that he was bedridden, even if he was elderly, and the story of the blazing thatch falling on him as he tried to enter his cottage would appear a distinct possibility, in which case he doubtless died of suffocation from the stinking smoke and fumes. It is also probable that old Miller, the sexton, and his companion, William Rabjohn, tried to pull the man clear and got themselves badly burned in removing the blazing thatch too late to save the millwright's life.

It was said that most of the houses destroyed in this fire were, at least to some extent, insured against fire damage but a subscription fund was set up to compensate those who were uninsured. The fund was only kept open for about one year and reached a total of less than £900. In present-day terms this seems a miserably small sum when it is understood that some fifty families were made homeless, but again, it was the loss of tools of trade that caused the greatest hardship until they could be replaced. All claims of up to £20 were quickly met on this account and shopkeepers, whose loss of uninsured stock threatened inconvenience for their customers, and might even have put the business at risk of closing, were compensated immediately funds became available.

A committee was formed at once to consider plans for rebuilding the devastated area. It was decided that the new Guild-hall should be built on a different site from that which the old one had occupied and a piece of land belonging to the Corporation was chosen and the work put in hand. At a specially convened meeting of the inhabitants of Bradninch the mayor's proposal that the old site of the Guild-hall and gaol should remain for ever an open space was adopted and this resolution has been observed to this day. All the arrangements for reconstructing the ruined streets of dwellings were, it is interesting to note, unlike the case of Chudleigh, as recounted in the previous chapter, carried out without the help of an Act of Parliament or its appointed Commissioners but entirely by the common-sense, determination, and goodwill of the inhabitants. Nor, as has been shown, were massive sums of aid sought, but enough only to meet the most pressing needs of the sufferers.

At the time of the fire of 1832, it appears likely that Bradninch's prestigous hotel, the Spread Eagle, was already past its greatest glory. There was, of course, no slackening of discipline among the numerous staff, no carelessness in the service, no reduction in the quality of the food and drink and the tankards gleamed as brightly, the barmaids smiled as politely and the welcome extended to all travellers was as warm as ever. The trouble was that new turnpike roads, which were inclined to follow the valleys and therefore to bypass towns on hilltops, like Bradninch, were taking away far too much of the 'passing' trade because the stage-coaches preferred them and the carriers, too. The turnpikes were generally a quicker way by virtue of their being far less hilly, but, what was of even greater importance to travellers, they had good new surfaces to ride over which did not break axle-trees or wheels as did the old, badly pot-holed roads. By 1835 the Spread Eagle had probably been compelled by falling trade to dispense with some of its staff and the number of coaches that rolled up to its door was almost certainly dwindling. Nevertheless, the grand hotel was still an object of local pride and merits further description. Apart from its unusual porch with the roof garden, already mentioned, the most important features of the house were the large reception rooms, two of which were on the ground floor and one, the great Assembly Room, on the first floor. The two downstairs rooms were divided by a splendid archway through which the horses were led into the fine stable-yard to be fed, watered and rested in the rows of stalls. In this yard, too, were ample coach-houses in which were chaises and gigs, and other hire-vehicles. The big, handsome room on the right of the archway was known as the 'Grand Parlour' and was for the convenience of travellers during their short stay.

At the head of a fine, broad staircase was the Assembly Room, which was used for dancing and, on special occasions, private balls given by the gentry. It was unlikely, however, to have been visited by the labouring people of the town but rather the tradesmen and professional men and their families.

The proprietor, Mr. Manley, was painstaking in his efforts to maintain the high standards of his house but, with some five other hostelries of varying size and quality in the town, competition must have been fierce and, no doubt, balls and public assemblies all too rare for the financial health of the inn.

The remuneration innkeepers received for their services to His Majesty's Post Office, when attending to the needs of the Royal Mail coaches was, to say the least, miserly and the trade from such passengers as were carried by them insignificant, because the guard in charge of the coach was allowed only five minutes to change horses and resume the journey. When it is appreciated how complicated is the business of changing horses, this time limit appears quite remarkable yet it was not unknown, when the mail coach had met with some unexpected delay, for the change-over of the eight horses to be accomplished in one and a half minutes. Such efficiency was costly as it required the services of at least six men, all of whom had to be paid by the innkeeper.

The stage-coaches were a different matter. Their journeys were much more leisurely and a few minutes lost here or there to pick up or deposit a passenger who would not be hurried was of little consequence. Furthermore, when arriving at an inn at an appropriate hour the passengers would take a hot meal before resuming their journey and, if the hour was getting late, would put up for the night. Naturally it was the stage-coach that landlords were most ready to welcome for it caused no fuss, no rushing to save a few seconds. The guard did not keep chasing the passengers back aboard with loud threats of leaving behind any who loitered, but allowed all transactions to go forward in a more civilized and unhurried manner, to the general benefit of all concerned.

The travelling 'gentleman' in his smartly turned-out curricle, too, was always a welcome visitor. He was sure to want his 'cattle' fed and groomed, he wanted a room in which he could wash himself and a good meal for him and his servant—for all of which he could be expected to pay generously. Unfortunately for inns like the Spread Eagle, he too was becoming a rarity, for the turnpike road, smooth and often straight, gave him a splendid opportunity to give his tandem horses, usually well bred, their heads and send them flying down the road.

The winter of 1835–6 was hard and did little to encourage anyone to venture further than absolutely necessary. The roads were icy through most of December and a searing, piercing wind blew constantly from the east. The horses drawing the mail coaches reached their changing points with rime-encrusted coats and had to be given extra attention in warm stables. The New Year saw no improvement in the weather conditions in North Devon and a few days into February the sky took on the appearance of a vast contusion, black and angry with streaks of yellowish-grey that seemed to remain stationary in spite of a rising wind that swung towards north-west. On the 9th, the snow began to fall in great flakes and continued to do so for several days; the wind blew in fierce gusts and drifting on a massive scale occurred.

Farmers called out their labourers and, armed with shovels and hurricane lanthorns, attempted the impossible task of digging out their sheep. Hundreds of ewes died, frozen to death as they stood together, heads to windward, behind some hedge or stone wall against which the covering snow drifted deeper and deeper.

On a Sunday night an old man named Peter Warren, who lived in a cottage at a little distance from the Spread Eagle, had his sleep disturbed by an unusual sound. It was as if thick ice was cracking and falling in large lumps from a height. He rose from his bed and shuffled across to the window and saw that the Spread Eagle was on fire. He managed to raise the alarm but the building was already well alight and even hot enough to have started to melt the snow and ice on the roof.

Staff and neighbours were soon astir and many people, wrapped in their warmest attire, hurried to the inn, but there was nothing to be done. It was useless to attempt to send a messenger to get a fire-engine because, even had a machine succeeded in reaching the spot through the deep snow, there would have been no water for its use, every well or pond being frozen to a depth of several inches.

Within an hour the pride of Bradninch, together with a row

of dwellings which had stood between the Spread Eagle and the old iron gates of the church, were completely destroyed. It was the end of what had once been the town's social centre and a landmark known to all who travelled that way. The inn was never rebuilt. Its glory, once the proud boast of the people of Bradninch, gradually became a faded memory and, to some extent, Bradninch faded with it into just another small Devon town, with inhabitants struggling to make a living the best way they could.

There were, however, advantages to be wrested even from such a grievous loss for those in a position to take them. When the rubble from the burnt-out houses had been cleared away and the land pegged out for rebuilding to begin, the owners of the plots got together and, under cover of darkness one spring night, shifted the marker pegs eastwards so as to take in a portion of the public highway which happened to be wider at the place than anywhere else in its entire length. When this piece of sharp practice was discovered, feeling ran high in the town, with much quoting of Mosaic Law regarding the moving of landmarks in order to rob a neighbour. The encroachers, however, brazened it out somehow and the new houses were erected, it might be said, partly on land stolen from the public by stealth.

In 1855 another fire, this time of short duration but swift destruction, consumed another sixteen houses at the east end of Bradninch. It started, as seems so often to have happened, in a bakery, on this occasion in one belonging to a Mr. Coleridge. Once again the ruins were quickly cleared away and the shops and houses rebuilt. There was no fire-engine in the town but it seems likely that this fire caused the people of Bradnich to get one. What they failed to get, however, was a better supply of water for its use. There was abundant water on nearby land but it belonged to the Duchy of Cornwall and the tenants to whom the land was leased were reluctant to allow any water to be drawn off, on the grounds that they needed every drop for irrigation, and negotiations came to nothing.

Sunday was to prove, yet again, an unlucky day for some Bradninch residents. In the small hours of that day, 4th May, in the year 1873 a fire started in the bakehouse of a man named Tucker. On this occasion the fire was not due to raked embers sending sparks up the chimney to ignite a thatched roof, for the day's work had long since ended and the baker and his family were asleep in their beds. The house next door to the bakery was occupied by the local carrier, a man called Mortimer, who was wakened by the noise of his restless horses in the stable behind his house. Surprised by the noise at such an hour, Mortimer went to investigate and the moment he stepped outside his back door, he could smell burning and saw smoke billowing from the bakehouse. He raised the alarm but the ground floor of Tucker's house was well alight.

Mortimer's first consideration was for his horses and he managed to lead them to safety. Other neighbours were quickly on the scene bringing ladders to extricate the family from their bedroom windows.

The terrified women and girls had only time to snatch up some garment to pull over their night attire before the intense heat forced them to run to the open window where they were helped over the sill and down to the crowded street. The servant girl, last to hear the alarm, came to a bedroom window naked but managed to snatch a sheet from the bed and wind herself in it before descending the ladder only just in time. Before her feet were on the ground the floor upon which she had been standing but a moment before burst into flames which licked out of the windows and set the thatched roof ablaze. Burning particles from the roof caused the fire to spread at an alarming rate.

In the few minutes since the alarm had been raised the narrow street had rapidly begun to fill with people, all ready to lend a hand if required but primarily to watch the spectacle. They were, however, kept constantly on the move as roofs on all sides burst into flames which presently leapt across the street and set fire to the Butcher's Arms public house.

Interest and assistance were instantly concentrated on the pub and willing hands were soon hauling the barrels of beer and cider from the cellar into the street, where those who had helped to rescue the precious fluid guarded it carefully, determined that not a drop should be wasted.

Mortimer, the carrier, having removed his horses to a place of safety had to manhandle his heavy wagon out of reach of the flames: it was, apart from his horses, his most valuable asset and it was probably already partly loaded for an early start on the journey to Exeter the next morning. He was compelled, therefore, to let his own house and its contents burn.

As soon as the alarm was raised, men had run to where Bradninch's fire-engine was housed and dragged it to the scene. If, as seems likely, this machine had been purchased after the fire of some eighteen years earlier, it would appear that it had not been cared for nor maintained in any reasonable way. Although water was still scarce, there was some at hand and attempts were made to use it but the would-be firemen were up against more than one difficulty. To begin with, they were unable to get the fire-engine close to the burning houses because their occupants had piled all their furniture, and other belongings, along the very middle of the narrow street and refused either to remove them or to leave them unguarded.

The *Exeter Flying Post* reported the performance of the Bradninch fire-engine thus: 'The town engine was speedily brought up and was found to be in a sadly shattered state and the hose ditto. More water squirted from the rents than was poured onto the fire, in fact, in a couple of hours it was entirely done up.'

As a result of the inadequacy of the machine a messenger was despatched to Cullompton for assistance but here again there was unnecessary delay, because the messenger was unable to persuade anyone to lend him a horse on which to ride the two and a half miles. It might be supposed that Mortimer would have been ready to lend one of his horses since his own house was on fire, but they would have been heavy draughthorses and would have had to go round by the highway, while a young man could have gone by a short cut over the fields in half the time on foot.

Meanwhile, thirst engendered by the smoke and heat of the blaze was slaked in the street outside the Butcher's Arms as barrel after barrel was broached by those who had rescued them from the burning pub. It was not until almost 7 o'clock, some four hours after the fire was first discovered, that the fire-engines from Cullompton arrived on the scene.

Before this help reached Bradninch some men had removed the thatch from some houses in the path of the blaze and had stopped the spread. The small quantity of water available to the engines was used to put out the smouldering remains of the sixteen houses and shops which had been destroyed and by about 7.30 on that pleasantly sunny Sunday morning many an on-looker began to make his tipsy way home to breakfast with plenty of gossip for his wife and possibly a disinclination to go with her to church.

There were no casualties in this fire and most of the sufferers were insured, at least partly, and there seems to have been no subscription fund started although some tradesmen lost everything they had. Perhaps it was a sign that people were becoming more independent of outside charity and able to rely on neighbours for a degree of help in such times of need.

The earliest record of a fire in Bradninch dates back, like the Great Fire of London, to the year 1666, when only the church escaped destruction. At that time it is probable that almost every building in the town would have been of timber construction. With a strong wind blowing and the fire starting on the windward edge of the town the streets would have been far too narrow to act anywhere as a fire-break. On that occasion most of the terrified population ran to the church for refuge, and, while the women and children, the elderly and the sick, stayed inside, the able-bodied men would doubtless have stayed in the street or the churchyard ready to beat out any burning debris that threatened to ignite any part of the fabric. After the fire the town was gradually reconstructed, probably to an entirely different plan from the old one, with more stone and cob used, but most of the roofs were again thatched.

The lintel over Mr. Mortimer's door had the date 1666 carved thereon, and the nature of his house and the adjoining houses was said to have given them a specially ancient appearance as they were among the first houses built after the Great Fire.

It will be observed, therefore, that, with 48 houses destroyed in 1832, a further unspecified number with the Spread Eagle in 1836, a further 15 in 1855, together with the odd single house fires in between, the blaze of 1873 which burnt down another 15 shops and houses and again, in 1884, 9 of the old houses in New Street were gutted by fire and, in 1897, on August Bank Holiday, the King William Inn and 3 other houses were destroyed, few if any of the houses rebuilt after the 1666 fire could have survived.

4

NEWTON ST. CYRES

During the 19th century, before brick buildings had replaced the old timber and cob, and tiles or slates had been substituted for straw thatched roofs, fire was looked upon by the small town dwellers of Devon as an unavoidable hazard in the face of which it was reasonable and sensible to abandon the burning dwelling to its fate while escaping, as fast as possible, with one's life and, if exceptionally fortunate, with a few cherished possessions. In the case of an outbreak of fire it was also accepted practice for the inhabitants of neighbouring towns and villages to despatch help as quickly as possible, initially with men and apparatus to fight the blaze and then with whatever necessities the victims required.

There is on record, however, an occasion upon which such assistance was flatly refused. It is a matter of speculation as to the cause of the original quarrel which led to what appears to have been an act of indiscretion, a display of bad manners and unwarrantable ingratitude on the part of the people of the little village of Newton St. Cyres. On the night of Friday, 1st May 1829, a fire broke out in the centre of the village and was, at first, blamed upon a widow who, it was thought, had carelessly thrown hot embers from her grate into a wooden shed behind her house. The widow, named Mrs. Pollard, occupied the second house in a block, or terrace of four, all of which were constructed of timber with roofs of the usual straw thatch. As was her custom, she had taken out the cinders from her fireplace and deposited them in the shed, believing them to be cold. By the time Mrs. Pollard had been two hours in her bed, that is to say at about 11 o'clock, the low thatch at the rear of her house was well alight and, within a few minutes had spread to the roof of her next-door neighbour, a shoemaker named Thomas Elliott. Shortly afterwards it spread in the other direction and ignited the thatch on the roof of Mary Davie's house.

The shoemaker seems to have been the first alerted, possibly by smoke blowing into his bedroom, and he quickly raised the alarm. Considering that it was the evening of May Day in a country village it seems a little strange that there were no revellers about the streets of Newton St. Cyres to have spotted the burning thatch sooner. Perhaps those wanting to celebrate the day had either gone the four and a half miles into the great city of Exeter or the shorter distance of three and a half miles into Crediton.

There was a light but fitful breeze blowing from the north-west that night, not a high wind but strong enough to carry burning fragments of straw across the road so that when the parish fire-engine, a small handcart type of applicance about the size of a perambulator, was brought to the spot it was used to keep the thatch wet on the cottages situated opposite the burning houses across the street, in the hope it would not ignite.

Despite a prolonged spell of dry weather throughout April there was, nevertheless, no shortage of water on this occasion as a small stream, or leat, ran along the side of the road in front of the houses, across the street from those on fire. The stream of water such an appliance would have been capable of ejecting would have been of little use in quenching the blaze in the wooden houses which were soon burning furiously.

There were a number of large residences in the vicinity of the village occupied by members of the landed gentry who, upon hearing the alarm, hastened to the spot and rendered whatever assistance they could, even taking their turn at the pump on the engine or organizing efforts to confine the fire to the four cottages which stood in the very centre of the village, so that every deviation in the wind's direction placed one building or another in jeopardy. One of the gentlemen, a Mr. Quicke, instantly recognized the danger to other property and the limitations of the village's own fire-fighting appliance. He therefore despatched a messenger on horseback to Crediton and another to Exeter.

The fourth house in the block on fire was occupied by a man whose name, like that of the shoemaker from the other end, was Elliott, his Christian name being William. His house was the last of the four to catch fire. By acting quickly, with the willing assistance of some young men, he contrived to remove to safety all the goods and chattels he possessed. Every effort was also made by these neighbours to save what they could from the houses of Mary Davie as well as Mrs. Pollard but the old, dry timber burnt with such fury and gave off such intense heat that it soon became quite impossible to enter them and those who were toiling in the street outside to prevent the blaze spreading must have feared being roasted and probably had to resort to spraying themselves occasionally with the pump.

It was, no doubt rather surprising to the people of Newton St. Cyres that the first fire-engines from outside to arrive on the scene were those of the West of England and the Sun Fire Insurance offices from Exeter, both of which reached the village soon after 2.00 a.m. In spite of the more powerful water-jets of the big machines the sturdy old timbers of the houses continued burning until about 7 o'clock.

When the large fire-engines arrived the blaze was at its height and for some time it was impossible, in spite of the efforts of the firemen, to prevent the thatch on the cottages opposite from igniting and it had to be stripped off and carted away to a safe distance. A little further up the street the Crown and Sceptre Inn was in grave danger when its roof caught alight but, by quickly removing the burning thatch and pouring on a great deal of water, the fire-fighters ultimately saved the building, with comparatively little damage done. From its doors went a succession of potmen and barmaids through the billowing smoke with tankards of beer to slake the thirst of the fire-fighters, the fire thus providing its own May Day jollifications.

The messenger who had been sent to Crediton to ask for the town's fire-engines to come to the aid of the village returned about the time the Exeter machines arrived and reported that his request had been met with a flat refusal, albeit with profuse apologies, while Crediton's two fire-engines stood idle in their shed, gleaming brass, polished woodwork and all.

The *Exeter Flying Post*, in its issue dated 7th May, 1829, states: 'We are glad to be able to state that the damage is confined to the cottages which are insured in the West of England and the Norwich Union Offices'. The report goes on to praise the untiring work of the members of the gentry who had hastened to the scene with an especial mention of the vicar, the

Rev. G. Carwithen, and adds: 'it was generally circulated that the cottages had been wilfully set on fire but we are creditably informed that such is not the case. It would appear that some wood ashes had been rather carelessly thrown into a shed at the back of Mrs. Pollard's cottage which, being still on fire, communicated to the woodwork and finally to the thatch of the building.' Where fires are concerned, the newspapers of the time seem to be interested mainly in establishing whether the burnt buildings had been insured by their owners. The *Flying Post* notes that three of the occupiers had taken the precaution to insure their goods and chattels. This apparent obsession is, perhaps, another indication of the attitude towards fires as inevitable hazards. With a note of admonition, the paper adds: 'Mr. Thomas Elliott, we regret to say, is uninsured.'

Having thus roundly blamed the widow, Mrs. Pollard, for starting the fire by an act of carelessness the *Exeter Flying Post* in its edition of 14th May, printed, without apology, the following: 'Whereas on the night of Friday the 1st of May, instant, the village of Newton St. Cyres was wilfully and maliciously set fire to. Notice is hereby given that a reward of £100 will be paid on the conviction of the offender or offenders on application to the Churchwardens and Overseers of the Parish and His Majesty's free pardon will be given to any accomplice provided he be not the party who actually set fire to the said village. Dated 11th May 1829.'

The odium of having started the fire was removed by the newspaper notice from Mrs. Pollard and put fairly and squarely on 'some person or persons unknown', and if it did nothing else it must have made her a little less unpopular with her immediate neighbours. What, then, was the evidence for such a downright statement? *Woolmer's Exeter and Plymouth Gazette*, in the issue of May 9th, had no doubt about it. It said: 'There can be no doubt that the fire was the act of an incendiary . . .' and repeats the story that it had, at first, been attributed to carelessness with hot embers, and goes on: 'This was proved to be impossible as the fire must have originated in the thatch. The woodwork over the ashes (deposited in the shed) not having been consumed. On Wednesday, a man going to an outhouse adjoining two of the cottages discovered a bundle of matches and other combustibles evidently half burnt and probably extinguished by the miscreant using too much force in thrusting the matches under the thatch.' After this report of a knowing piece of detection the paper went on: 'A man has been apprehended on suspicion and the Secretary of State has been informed of all the circumstances.' The report ends on a lighter note: 'The Vicar, The Rev. G. Carwithen, was evidently most impressed by the unstinted exertions of his poorer parishioners, as well as those of the gentry who were present and on Saturday morning regaled those who assisted at the fire with as much bread and cheese and "good old Devonshire cider" as they could conveniently make away with.'

It may be imagined that while they were partaking of this refreshment conversation must have turned to the matter of Crediton's refusal to send their fire-engines when requested. There may have been some who were not altogether surprised by this apparently heartless response. Further light is thrown on this unusual attitude towards a smaller neighbour by a letter which was published in the pages of the *Exeter Flying Post* of May 14th. The letter seems to place the blame squarely on the inhabitants of Newton St. Cyres. It reads:

> Sir,
>
> It is with extreme regret that I noticed in your paper of Thursday last a paragraph by which it appears that a correspondent imputes a great blame to those who have the custody of the Fire Engines at Crediton, on account of their refusing to let them be taken to assist in the extinguishing of a fire at Newton St. Cyres. It certainly is not to be wondered at that your correspondent complains in severe terms upon the subject as it is natural for sufferers on such unfortunate occasions to look for aid amongst their neighbours but is it not right that an individual, who is perfectly innocent of any uncharitable feelings towards the sufferers, should bear the odium of such refusal and I am sure that the inhabitants generally of Crediton and its vicinities will bear me out in saying that a man less deserving of such opprobrium than the individual to whom the application was made does not exist and who expressed at the time his deep regrets and that he was so situated as

to compel him to do an act so revolting to his natural feelings. The individual I allude to is Mr. Abraham Wreford of this place and I believe that everyone who knows him well will readily acquit him of any unkind or uncharitable intentions and I trust your publication of the following statement of the facts will convince your correspondent, and the public, that he could not have acted otherwise.

The Fire Engines at Crediton are the sole property of the inhabitants of the town, having been originally purchased by subscriptions raised for that purpose. Some years ago a fire broke out at a neighbouring village and the Crediton engines, on application of the inhabitants were immediately sent to their assistance. In this service they were materially injured and the repairs amounting to between £20 and £30 were borne by subscription by the inhabitants of Crediton. As the people to whose assistance they had been sent not only refused to pay a shilling towards it but they also refused to take the engines back to Crediton and suffered those of the latter place to fetch them themselves. In consequence of this, at a subsequent meeting a committee was appointed for the purpose of keeping the engines in complete repair and rules and regulations were formed to be observed by them and by those with the care and the keys of the Engine Houses. The following is a copy of the rules applicable to this case. Rule 5. That in the case of a fire happening in any part of the Parish of Crediton beyond the vicinity of the town, and within the limits herein after set forth, not more than one engine to be permitted to leave the town at the same time, the other to be retained in the town for the protection thereof. Rule 6. That the distance on the several roads or avenues leading from any part of the town is not to exceed two-miles-and-a-half, namely from the milestone on the hill to the aforesaid distance on or near the Exeter and the Fordton roads: from the West end of the town to the same distance on the Westward Dunston and Barnstaple crossroads, or to any other place to the same distance within the Parish of Crediton. Having finished this statement of fact I must leave it to the public

> to determine whether the inhabitants of Crediton were warranted in coming to such resolutions or not but that such resolutions were passed and that the custody of the engines were entrusted to Mr. Wreford under strict directions to abide by them in all cases is quite certain.

The letter is signed 'A Constant Reader' and is dated 12th May 1829.

5

NORTH TAWTON

The pleasantly quiet small town of North Tawton is beautifully situated on the east side of the green valley of the River Taw. Here, the river flows silently and, unlike most of North Devon's inland waters, seems to have no urgent desire to reach the sea. On a warm summer afternoon its tranquil dignity seems to set the mood for the town, which is some 10 miles west of Crediton and a little more than 6 miles north-east of Okehampton. It stands on an ancient site where dwellings have stood since long before Domesday Book was compiled, for it is probable that people dwelt here well before the Romans came to Britain.

Like so many of its neighbouring towns, North Tawton once had a prosperous woollen-manufacturing industry together with an important tannery. It is likely that the woollen mill and the cottage weavers accounted for more than 50 looms in the town as late as 1835, at which time it was a thriving market town, busy and noisy and with a population of about 1800 inhabitants. They occupied about 270 houses of one kind or another but mostly of the usual Devon cob covered with a straw thatch roof. The town also boasted a few good stone-built properties, such as its ancient coaching-house, the Goswick Arms, an inn sometime compared by an enthusiastic newspaper columnist with the Angel at Islington and the Bell at Edmonton (made famous by the poem about the ride of John Gilpin).

On Saturday the 23rd June 1832, the people of North Tawton were preparing a revel, a celebration of the passing of the Reform Bill. It was, of course, to be nothing so lavish as the great dinner being organized by the mayor and aldermen of Taunton, who a few days later were to regale 5000 people with a hot meal of roast beef and plum pudding.

On that Saturday people from neighbouring towns were coming into North Tawton to plan their part in the revel and rehearse whatever they intended to do. It may be imagined that there would have been fiddlers from Hatherleigh, mummers from Okehampton and Morrismen from Winkleigh Town. Nor would the housewives of the town be idle on such an occasion. Their ovens well stoked to generate a goodly heat, many of them would surely have been hard at work, their arms white with flour up to the elbows as they mixed and beat, cut and rolled pastry and cake mixtures on the scrubbed tops of large kitchen tables in readiness for tea parties in the street.

The little town must have been in a state of excitement with children running here and there peeping at what they were not supposed to see until the morrow while the strains of the fiddles and hurdygurdies would have been heard as some lively tune was practised, some intricate dance rehearsed. In the midst of all this feverish preparation there suddenly came the cry of 'Fire!'

It happened at about 3 o'clock in the afternoon when all the townsfolk were preoccupied with a multitude of chores and at first there may even have been doubt in the minds of some as to the cry being genuine and not a boy's prank. The reality of the alarm was soon apparent as a special newspaper correspondent, an actual eyewitness, wrote the next day for publication in the *Western Luminary*. It reads as follows:

> In the afternoon of yesterday, on the eve of our Revel, a fire broke out here whose devouring fury the greatest efforts were unable to check until, by its ravages it had rendered our village, before happy in anticipating the mirth of the morrow, into a scene of desolation which can only be appreciated by eyewitnesses. By this distressing calamity between 20 and 30 houses have been burnt to the ground and as many as 40 poor families have been deprived of a home and so rapid was the progress of the fire that few were able to save even a bed from the destructive fury of the element. We are happy to be able to add that no lives were lost.

The records show that the total number of houses destroyed that afternoon was 30 and the cause of the outbreak was unknown and therefore almost certainly accidental. It was once again the old enemy of the thatched cottage: a long spell of dry weather and a brisk breeze with a few stray sparks from a chimney. It may have been caused by the stoking of an oven for the purpose already described or it could just as easily have happened when someone wanted to make themselves a cup of tea for it has to be remembered that in order to boil a pan or kettle a fire had to be kindled on the hearth. Thus the risk of fire from sparks was ever present and once the thatch had caught fire the wind, as in this instance, spread the flames with alarming rapidity.

That sunny June afternoon the fire blazed so fiercely that, although there were willing hands in plenty to help by fetching water, it quickly spread from cottage to cottage, setting light to many outhouses, stables, pigsties and sheds as it went. Many of the women left whatever they were doing and hurriedly carried furniture from their houses to wherever they thought it might be safe while men pulled down houses in the path of the flames to make a fire-break, only to discover that the wind had carried burning straw to the cottages beyond those they had demolished.

In these dreadful circumstances thoughts of revels were forgotten and men risked their lives in an attempt to stop the spread of destruction and to help neighbours salvage their belongings.

News of the disaster was conveyed to neighbouring towns and two fire-engines were hurriedly despatched from Okehampton and another from Crediton, all three arriving in North Tawton in what the *Exeter Flying Post* described as a remarkably short time 'considering the distance'. In spite of the valiant efforts of the firemen it was several hours before the fire blaze was subdued after which many a disappointed would-be reveller wended his way home to spread the unhappy news that the celebrations, so eagerly anticipated, would not take place. No doubt this decision provoked angry opposition from those inhabitants whose homes had not suffered. There seems little doubt that from this time an undertow of antagonism seemed to grow between the inhabitants of different parts of the town; an antipathy which was to come to light two years later. Although the majority of the dwellings destroyed that afternoon housed poor people, the landlords, in rebuilding, were long-sighted enough to use bricks and slates for the replacement cottages and the occupants were no longer subject to the constant danger from sparks falling from the chimneys on to their roofs. Perhaps this fact may have caused jealousy among those with straw-covered houses. In spite of this calamity, or possibly because only a small part of their town was burnt down, the people of North Tawton were not moved to invest in a fire-engine of their own but were evidently content to rely on their friendly neighbours. This was to prove a miscalculation which would cost them dear. Furthermore, the fire which caused the revel to be abandoned evidently gave rise to second thoughts among the inhabitants who probably realized they had little to celebrate in the passing of the Reform Bill, as few of the labouring class would have met the requirements that would have entitled them to vote and there appears to be no record of the town celebrating the bill at a later date.

Two years after this sad occurrence another disastrous fire broke out in North Tawton. In this fire, which happened on Monday 14th July 1834, more than 60 houses were gutted. About midnight a man saw that the roof of a house was smoking from the back and upon going round to investigate discovered a dairy behind the house was well alight.

The wind was quite strong and gusting from a westerly direction and houses both up and down the street (now Fore Street) were taking fire one after another. At first the blaze was confined to one side only and, the alarm having been raised, people began appearing in the streets to render what help they could to confine the outbreak. *Woolmer's Exeter and Plymouth Gazette* stated: 'The greatest consternation prevailed among the inhabitants whose situation was most distressing being, as we are informed, without a Fire Engine which circumstances is extraordinary considering their late calamity'.

As had occurred when the previous fire broke out, someone was despatched to neighbouring towns to summon help and with commendable promptitude the Hatherleigh and Okehampton engines set out to assist, but because of the distance and the time it had taken the messenger to reach them and raise the alarm, it was almost 4 o'clock in the morning before the Hatherleigh appliance, drawn by a pair of horses, drove into the burning town.

During the interval between midnight, when the fire was first noticed, and 4 a.m., the inhabitants had made every possible effort within their power to stop the fire spreading but in the first hour some thirty houses had become one mass of flames, with crumbling cob walls exposing white-hot stones that soon rolled into the street threatening anyone foolish enough to get too close. Few attempted to approach these ruins through the dense, impenetrable smoke and the searing, blistering heat which the newspaper described as 'awfully grand'.

Unlike the fire of two years earlier, when the burning houses had been remote from the church, these blazing cottages were dangerously near that part of the town and soon after 1 o'clock the trees which surrounded the churchyard presented a terrifying spectacle to hundreds of people who had gathered therein with their salvaged possessions, when six of the great trees were seen to be on fire and, indeed, were soon blazing like giant torches.

The houses to one side of the church appeared to be absolutely secure as the wind was blowing in a contrary direction and a considerable quantity of shop stock, together with other valuables had been lodged in them soon after the fire began. The thatched roofs of these houses sloped down very low at the back and it was in one of these that another fire started without, apparently, the least cause or, as *Woolmer's* put it, '... Fires burst forth most suddenly and with a force which baffled every effort to subdue them to the indescribable grief and sorrow of the inhabitants who saw what they considered their only place of safety fast becoming a mass of ruins and much of the property collected therein, too.'

Again the fire seemed to burn with renewed fury, leaping from house to house until almost the entire eastern side of the town which had escaped the fire of 1832 now presented a dreadful spectacle of devastation.

Nor was the ancient parish church to escape unscathed. A gust of wind carried burning strands of straw to its shingled spire, which was soon burning. At one moment not only the venerable old steeple was threatened with destruction but even the old South door with its sanctuary ring. Men ran to do what they could to save the church but before they could scramble up to the spire the top was already well alight. They eventually succeeded in quenching the flames but not before the upper part had been severely damaged. Later when it came to repairing the damage there was insufficient money to restore it. To this day it has a sawn-off, stunted appearance which might be said to constitute a noble monument to the many acts of courage shown that night by the inhabitants of the little town.

As in the fire of two years earlier, houses were pulled down in an attempt to make a fire-break but it was not until 9 o'clock on Tuesday morning that the conflagration could be said to have been got completely under control. The Goswick Arms escaped the blaze but the rest of the houses in that part of the town were destroyed, including two other public houses, the Fountain Inn and the George, together with a church school and some 60 dwellings, several large barns and a number of stables and other outbuildings. One horse died, together with many pigs which people had kept in their back yards; also a great many chickens were burnt or, at any rate, disappeared from the premises of their rightful owners.

The newspaper goes on to say that much of the property destroyed was uninsured and that the total damage was estimated at £10,000.

In the morning people began running hither and thither seeking news of relatives and friends, many of whom were doing the same, in a grotesque game of hide-and-seek. At the height of the blaze the workhouse took fire and its inmates, the elderly, the sick, the mad, the halt and the blind were, with difficulty, removed and lodged in the safety of the church until some more convenient shelter could be found for them. Many people had lost all they had, some escaping from their burning cottages with no more than some scanty garment thrown over their night attire so that their first care, when daylight came, was to seek friends or relatives from whom they could beg or borrow adequate clothing. Suddenly, therefore, those whose houses had escaped the fire, faced with upwards of a hundred families in need of breakfast, were in a turmoil of uncertainty as to where to find extra food as at least one bakery had been destroyed together with its stock of flour.

It was remarkable that no life was lost in the fire and newspaper reports relate only one accident: that of a young man who suffered considerable bruising when he fell from the roof of a house while attempting to remove burning thatch.

The fields surrounding the town must have presented a strange sight with piles of furniture and household belongings all mixed up with horses, cows and pigs while their distressed owners stood guard over them hoping, no doubt, that the weather would remain fine until they could find temporary shelter. There is evidence to suggest, too, that goods left unattended in streets or fields would be at risk from some criminally inclined persons known to live in the town, as indeed, *Woolmer's Exeter and Plymouth Gazette* makes quite clear. The language the writer uses to express the matter is somewhat quaint but worth quoting as an example of what the reader of 1834 would have been accustomed to. It reads as follows:

> We deeply regret that in closing our account of this melancholy occurrence that we are under the necessity of stating that very strong suspicions are entertained that this destructive fire was the diabolical work of incendiaries. We sincerely hope, for the credit of human nature and for the honour of our County which has hitherto been sustained from this most shocking crime, that these suspicions will prove groundless but, looking at the unaccountable manner in which the houses near the Church took fire, and connecting this with other circumstances which have been related to us, we cannot help feeling that there are very strong grounds to show that the vile hand of the incendiary was engaged in this work of fearful devastation. We have heard that when the people of Hatherleigh, who came with all possible speed with their engine to assist in staying the progress of the flames arrived in the burning town, they were greeted by some heartless villains with stones and were otherwise insulted and that one man who had fixed a ladder against a house which had caught fire on the roof, on which he got in an endeavour to put it out, was left in that dangerous situation to get down in the best manner that he could, some scoundrel having taken away the ladder. It has also been stated that matches have been found in some of the thatch that was pulled down and that

> since the last fire (in 1832) some of the most debased charges on the Parish have been heard to say ''Tother end o' the town will be on fire next'. We are unwilling to give credence to all these reports but at the same time it appears that there is a set of immoral and desperate characters in the neighbourhood and we are glad to find that a rigid enquiry into the case has been instituted by the Authorities who are resolved to use every possible effort to trace the origins of this dreadful calamity.

As with most newspaper reports of the period, attention was given to insurance protection and, in this case, it was again suggested that too few people had insured their property and those who had done so included many who had underinsured. Later events were to show, however, that at least one family had taken out very adequate cover indeed.

From the reports on the 1834 fire at North Tawton, it appears that the enmity which had arisen between some of the residents in the west quarter of the town and whose houses were burned in 1832 and those who lived on the east side, was not due in any way to class distinction, and may have arisen because the victims of the earlier fire thought their neighbours on the other side of town had been less than generous in helping them. Whatever the cause there can be little doubt that antagonism existed between a small number of the inhabitants, and in such circumstances rumour spread as tongues wagged and fingers pointed until the authorities felt compelled to act and, after some enquiries, three people were arrested and charged with arson. One of them was John Taylor, a licensed hawker, and with him his wife Jane, and his father Richard; all were charged with the same offence and were taken to Exeter Prison to await their trial.

On Monday, 23rd March 1835 the three accused appeared before Judge Baron Gurney and a Grand Jury. The proceedings created considerable local interest and reporters from all the newspapers crowded into the court, where they quickly received a surprise. All three prisoners having pleaded 'not guilty', the court proceeded to dismiss the charge against Richard Taylor on the grounds that he had no case to answer.

Jane Taylor, aged 42, was charged with having wilfully set fire to a certain dwelling-house in the town of North Tawton on the morning of 15th July 1834 with intent to defraud the Norwich Union Fire and Life Insurance Company, and John, also aged 42, was similarly charged that, being the husband of Jane, he had incited and moved her to commit the said felony.

The prisoners appeared in the dock about 10 o'clock in the morning. The case was brought by the Parish of North Tawton and the indictments numbered 20 in all.

Mr. Tyrell opened the case for the town and having outlined the charges called the first prosecution witness, a Mr. Pleace, a labourer of North Tawton, who told the court, 'I know the prisoners at the bar. I lived about three or four doors from them on the opposite side of the street. I knew their house well having been in it but not for three or four months before the fire took place. I remember the fire on the morning of 15th July last. After I was in bed I was alarmed by a cry of "Fire!" I instantly got up and went out of my own house to John Turner's, the next door up the street from Taylor's. I went to the back of Turner's premises and observed a fire in a little sling (a lean-to outhouse). I believe he used to use it as a dairy when he kept cows. I believe there was no communication between the roof of the sling and the dwelling-house. There was no fireplace in the dairy. It was burning inside. The flame had not made its way through and I did not stay until the flame made its way out. I turned out Turner's pig and came back again to the front.'

Cross-examined, he said: 'I live farther from the church than Taylor. The dairy was thatched as were all the houses near. I went back again to move Turner's things.' In answer to a question put by the judge he said 'Between sixty and seventy houses were burnt. I believe it was about a quarter after twelve when I was disturbed by the cry of "Fire!"'

The next witness to take the stand was the Taylor's next-door neighbour, John Turner, who, having told the court where he had lived at the time of the fire, continued, in answer to Prosecution Counsel's question: 'John Taylor is a licensed hawker. Soon after midnight on the morning of Tuesday 15th July, last, I was awoke by a cry of "Fire!" A man named John White broke open my door. Jane Taylor, the prisoner, knocked at the partition that separated our houses and cried, "Fire!" two or three times. I don't know whether I heard John White at my door or the cry of Jane Taylor first. I began throwing my things out of the window.' Questioned further, Turner said: 'I had been in Taylor's house on Sunday morning and I did not see that anything was wanting in the house. Things appeared just as they used to do. He had not more furniture nor better in quality than my own only I think he had a better bed. I think mine was worth £24 or £25. Taylor's house had one room downstairs and one up. I had a conversation about six or eight weeks before the fire when he said if he were to insure he should not think of an amount less than £100. I don't know, but I don't believe he was at home that night. I saw the prisoner, Jane Taylor, about ten or eleven o'clock that morning in the churchyard and I had a conversation with her concerning what had happened. She said she first saw the fire burning up by her mother-in-law's bed. She lived in the house with them and his father also. She said her husband and his Father had gone to Winkleigh to buy a horse. I asked if he had gone with his pack or not. She said "No".'

Cross-examined by the counsel for the defence, he told the court: 'The partition between our houses was thin. I think I first heard the cry (of alarm) in the street but it could not have been but half a minute between that and Jane Taylor knocking at the partition and crying "Fire!" in her own room but I can't say distinctly which was first. I believe they all slept in the room upstairs when home. Jane Taylor's mother-in-law was not at home that night. I don't know what part of the room her bed stood in. She had been away five or six weeks.'

In answer to a further question Turner admitted that there was easy access to the rear of the houses through a hole in the hedge into Barton Lane and which was large enough for a man to pass through quite easily. The witness added: 'I known a man of the name of Ford who was apprehended on this charge.'

The judge then asked for further information concerning the furniture in Taylor's house to which Turner replied: 'There were two dressers in Taylor's kitchen. That belonging to John was worth I think about £8, the other belonged to Richard Taylor.'

Here the Judge recalled the previous witness, Pleace, and asked: 'When you were behind Turner's house and saw the fire in the sling house, or dairy, did you see any fire in Taylor's dwelling?'

'No' answered Pleace, 'the windows in Taylor's house look into the street. When I was going by I did not see any fire in Taylor's house but I saw smoke coming up above the roof. I could not see then whether it came from the roof of the dwelling-house or the attached sling house behind but when in the garden I saw the smoke coming up out between the two roofs exactly where the thatch was divided.'

Pressed upon this point by the judge, the witness described it as a slight break in the thatch where the two roofs joined. He continued: 'The fire was in the sling house, I think so. There was a window in the sling house towards the garden and I could see the fire through that window.'

The next witness, Arthur Heywood, said: 'I'm a labourer and live at Broadwood Kelly. I know the prisoner John Taylor, I remember meeting him before the fire at Broadwood Kelly in Reed's house. I remember a conversation which took place there. It might be twelve months before the last fire at North Tawton.' He had, he told the court, heard Taylor say that he knew a man who had insured his goods for £100 and his father's for £50 but reckoned them not worth more than £10 and Taylor said he made no doubt that the house would be '. . . set afire some time or other for 'twas insured in such amount above the value then the man would draw £150 from the insurance.'

In answer to a question by the Judge, Pleace said: 'I believe the prisoner was sober at the time'. Cross-examined, he agreed that the conversation had been about the fire at North Tawton in 1832. He was sure he had heard aright as Taylor had spoken loudly because Reed, a blacksmith, was hard of hearing, and he concluded: 'He spoke loud and openly.'

John White, who had been to call Turner on the morning of the fire, then gave his evidence. He said: 'I saw Jane Taylor while I was at Turner's door. She came out of her door and said, "The Lord have mercy upon me I've got no person to help me". She turned round, locked her door, went to the other side of the street and sat down upon some steps. She was not dressed. She had her under-petticoat on and her shift and her grandchild in her arms. I saw her again on the same day in the town and had a conversation with her. She said "I suppose you're going to be a witness against me about the fire?" I said, "What can I be a witness against for I didn't see you set it afire?" She said, "They say I set it afire". I said, "I didn't see you do it". She said, "I'm sure I did not". I told her I saw her lock her door but she didn't say anything to that.' He went on to tell the court he had stayed up late on the night of the fire because he was expecting his horse and cart to be brought home and was the first to see the fire. 'A lick of smoke came into my house—I saw it came from the back of John Taylor's house and I went and broke open Turner's door. I believe I was the first to give an alarm but a man rode through the town on horseback and I asked him to give an alarm as he rode along.'

Giving further evidence, White said he knew Taylor's house well and described an oven in the downstairs room against the wall beyond which stood the sling house. He said he took Turner's children out of his house to his own dwelling across the street. Asked about Jane Taylor's grandchild, he answered: 'Jane Taylor kept her grandchild always when she was at home. It is the child of Thomas and Loveday Hooper. The child was wrapped in a blanket.' He had seen nobody before he knocked at Turner's door. He also said he did not know the direction in which the wind was blowing but said the smoke blew over his house.

At this point proceedings were held up while the judge, counsel for both parties and witness discussed the direction from which the wind was blowing when the fire broke out, after which it was generally accepted that it blew from the south-west.

White continued his evidence: 'I don't know how soon any fire broke out in the churchyard,' he said, 'but it did break out there. My house was burnt down.' In answer to a question from the Judge he said, 'When Jane Taylor locked her door I didn't see whether she took out the key or not. The fire went from house to house but I didn't see it burn out in the churchyard. I know Lock's house (near there).' White then stood down.

Evidence was then given to establish where John Taylor had been during the time and fire had burned and a woman witness, Ann Gillard, was called, who told the court that she resided at

South Tawton and knew the prisoner, John Taylor. 'I remember him being at my house about six weeks before the fire. He had said he was not coming into the place but once more and then was not going to take his pack (goods for sale) with him. He said, so help him God, he had not taken but threepence since he had been in the place (South Tawton). He said such parishes as North Tawton and South Tawton didn't ought to be let stand, they ought to be burnt down and that man who put it out the last time it was afire ought to have been thrown into the blaze, be damned if he didn't.'

Cross-examined, Ann Gillard said South Tawton was six miles from North Tawton and Taylor had come there with his pack. 'I bought some things off him but that was after he hadn't taken but threepence.'

The next witness was a young man named John Hearding, who told the court that he remembered well the last fire in the town. 'I saw Jane Taylor sitting under her daughter's window with a child in her arms and a blanket over it. I saw her again about half an hour afterwards when she came to the churchyard. She had the child wrapped as before. I was there watching over my mother's goods. She (Jane Taylor) sat down by her daughter's things. She hadn't any bonnet and I asked her if she had any clothes on. She said "No, nothing but what I saw." I asked where her husband and his father were last night and she said they had gone to Winkleigh. I then asked if he (her husband) had taken any goods with him. She said he had not taken out any with him. I asked where his pack was and she said "Burnt with the rest of the things". I didn't hear her ask for any assistance.'

Cross-examined, Hearding said: 'At the time I first saw her at the churchyard the fire was still burning. Everybody was bringing goods to the churchyard. I was there when the fire broke out at Mr Lock's house, that's where the fire at the churchyard began.'

The evidence, which appears to have been presented in a somewhat haphazard manner, then reverted to the behaviour of Jane Taylor, and William Brealey next gave evidence. He said he was a shoemaker living at North Tawton. 'In the night,' he told the court, 'I was alarmed by the fire and left my house. It was somewhere between midnight and one o'clock. I went to John

Taylor's and tried to open the door but it appeared to be locked or bolted and I did not open it. I didn't find any key in the door.' Cross-examined, he said the door was latched, and he did not see a key in the door, and added, 'I hove the latch but it would not open and I was in a hurry.'

It seems a strange thing to have locked the door of a burning house and to have abandoned all her possessions to the flames. Perhaps Jane Taylor, alone in the house, had been frightened out of her wits and was not quite responsible for her action yet none of the testimonies of the witnesses suggest that she showed any sign of fear.

Another neighbour, Sybella Pike, was then called and said: 'I know the prisoners' house. I saw some goods going away from it about six weeks before the fire. I can't say what goods they were, 'twas boxes. I saw some going away another time. They said they belonged to Jane Bradford, a daughter of old Taylor.' Cross-examined, she admitted that she couldn't be sure Jane Taylor told her so. She went on, 'I was in the habit of going into the house but I haven't been upstairs for, I supposed, three months before the fire. I didn't consider their goods worth more than £20 but I don't know what was in their boxes or what bedding they had. I know nothing of his pattern goods.' Further cross-examined, she replied: 'My house was burnt, I might save one half but it was ransacked about.'

Once more the evidence turns to John Taylor and the next witness called was a farmer named James Hammett who resided in the parish of Winkleigh, about three miles from North Tawton. He said he remembered the fire at North Tawton and continued: 'I didn't see John Taylor myself at my house the day before the fire, the Monday, but a pack was left at my house that day. Taylor owned that pack to be his. I carried it to North Tawton before the Justices. John Taylor examined it there before them and acknowledged it to be his. That was on the 29th of July. I saw Taylor on the 19th July at my house and his wife was with him. He didn't take his pack off that day. He said he was thinking of taking a shawl from the pack for his wife but he didn't. We had a conversation about the fire. He said it was distressing, and things of that kind, and he told me he had been examined by the agent to the insurance office and that he had given in an account of his loss. He did not tell me the amount. He showed me the receipts of the insurance which he had paid and I think took out the policy but I didn't see it. It wasn't opened. He said he had paid two years' insurance. He produced a handful of sovereigns, 20 or 30 perhaps and said he was doubtful about getting the insurance but didn't want for money. I think he said his wife had some money in her pocket, too, or something of that sort but didn't say much.'

Cross-examined, Hammett told the court that 'Monday was Winkleigh Fair Day. I saw the prisoner at the Fair but not at my house that day but he has been there on different occasions. I never saw his wife until the 19th July.'

Roger Wilcocks, who seems to have been some sort of minor official of the parish of North Tawton, then gave evidence regarding the travelling pack which had been left by Taylor at Farmer Hammett's and subsequently delivered by him to the local justices. 'It was delivered to me,' he said, 'by order of the magistrates when Taylor was committed. Just when he was being taken away he said to the Hon. Newton Fellowes that he hoped he would have mercy upon him.'

Another switch in the direction of the evidence brought a young lad to the witness box whose name was Simon Bickell. He told the Court that he remembered the fire on the night of July 14/15th at North Tawton and said he was in the churchyard that morning. 'I saw the prisoner, John Taylor, there. I can't say what time it was but I was in the churchyard by one o'clock, I think it was getting light when I saw him, it was light enough so that I could see him. I knew him before and had a sufficient view to know it was him.'

Cross-examined, the lad said, 'I am fourteen or fifteen years old. There was a great many people in the churchyard and I stayed there most of the night. I was with our goods. Taylor was standing still and Richard Taylor was with him. There were five or six people round him. I did not hear whether they were talking or not: John Taylor had on a dark frock-coat, whether blue or black I can't say, white stockings with short breeches. He was never near the lower church gate near Lock's house.'

Was the boy's evidence that he had actually seen Taylor in the churchyard while the fire was still blazing, reliable? His description of the hawker's attire sounds somewhat suspect. Such a man would have been likely to dress less formally for carrying a heavy pack on his shoulders. The following witness also seems to give the lie to the boy's story, another farmer, whose name was Jeremiah Harris, who said he lived near the village of Ashreigny. 'I know the prisoner, John Taylor, I saw him the same morning as the fire at 11 o'clock in the forenoon. He was going towards Winkleigh Raceground, that was in the direction towards North Tawton, about five miles from it. I gave him the time o'day and asked him whether he had been home to North Tawton tonight. He said he had not. I then asked him if he had heard that North Tawton was burnt down. He said "No". We then parted, he towards the Racecourse and I to Winkleigh Town. The way he took was also the way to North Tawton as I have heard. He had come from the direction of Winkleigh. He made no remark whatever when I told him North Tawton was burnt down.'

Replying to the judge, Harris said, 'I heard of the fire about nine o'clock, it was the common talk all through Winkleigh Town.'

At this point we are reminded that one charge brought against the prisoners at the bar was attempting to defraud the Norwich Union Fire and Life Insurance Company, and John Ellis, Senior, local agent for the insurance company, went into the witness box. He told the Court that he had arranged the effecting insurances with John Taylor some time after Michaelmas 1832 for £40 for goods and furniture, £10 wearing apparel and £50 stock-in-trade of drapery goods and also, in May 1833, by his father for £40 household furniture and £10 tools. Ellis also stated that there were no other instances of a father and son being insured with him for these amounts. The claims made were understood to be, for John Taylor, £42 19s. 0d. for goods and furniture, £6 10s. 6d. for wearing apparel and £53 0s. 3d. for drapery goods. For Richard Taylor a total claim of £51 1s. 2d.

'When he effected his insurance with me,' said the witness, 'he said he had drapery goods generally to the amount he insured of £100. I saw him after the fire took place when he came to me and enquired when Mr. Hartnoll, the agent at Exeter, would come down to survey the damage, and to have his claims allowed.'

Here the claims document was handed to the judge and Ellis continued: 'Mr. Hartnoll came down. He told Taylor he didn't think his claim would be allowed. He (Taylor) turned rather sulky on this and said he did not care a damn whether they were or not. He supposed the next fire must be in the open fields.' Suggesting, perhaps, that if the insurance companies dodged all their obligations there might be no rebuilding done.

Here the defence seem to have brought a witness. He was William Green who had formerly been a clerk in the Norwich Union Office, but was then employed in the Stamp Office. He verified the signatures of the directors attached to the policies which were now put in.

The Exeter agent, Thomas Hartnoll, then gave evidence. 'The claims,' he said, 'were first brought to me by Mr. Ellis, but I afterwards saw both the prisoner at the bar and his father upon them. My suspicions were excited finding these two claims only made out when I came down to North Tawton on Friday (following the fire on the Tuesday).' Witness pointed out that the claims exceeded the amount insured in each case. Here the Judge reminded the jury that the case was not being brought by the Norwich Union but by the North Tawton authorities.

Then John Ellis, Junior, took the witness box and gave this evidence: 'I remember the fire at North Tawton and I saw the prisoner, John Taylor, on that day, at about midday. I asked him if he had saved any goods and he said he had not. He had not been at home and that Jane took the child and when she came out locked the door and took out the key and went to call Loveday (their daughter). Between then and the evening I saw him again. He asked me if he had any call to see my father about the insurance money. I told him there was no hurry for that. I saw him a third time the same day when he asked me if I thought he would get his insurance money or not. I said yes, if everything was right he'd be sure to get it. He said if he didn't get it it would be no matter. I told him he would have to make out an inventory of all he had lost.'

Cross-examined, Ellis said: 'I assisted him in making out the inventory of his goods. He had asked me to do so on the Thursday after the fire but I refused to do so at first. I wrote it from a pencil copy which he brought. I wrote no claims for any other persons. I was at the churchyard on the night of the fire at different times. I did not see John Taylor there.'

Here the judge intervened to recall the witness John Turner, next-door neighbour of the prisoners before the fire. He told the court that the room in which the Taylor family slept had no ceiling, it being open to the thatch of the roof which slanted from front to back of the house. There was, he said, a party wall between their kitchen, or downstairs room, and the sling house or dairy. 'The upstairs was plastered at the sides about three feet up. The fire must have come in above the laths. Gillow's house was below Taylor's and Mollie Moore's bakehouse was behind Skinner's. The chimney stood in the middle of the bakehouse and came out of the top in the middle of the thatch of the house.' Further questioned, Turner said: 'The ovens were heated with furze and there was a hole in Taylor's outhouse with Skinner's which was stuffed up with furze. It was, I suppose, eight or nine feet from the ground and just under the thatch.'

The Judge then asked: 'Was Mollie Moore's bakehouse in use?'

Turner replied: 'It was used constantly twice a day.'

This closed the case for the prosecution, to which the prisoners had put in a spirited defence, denying, in the strongest terms, the charge, the commission of which they declared they had no inducement, their property being more than equal to the amount they had insured for. They drew attention to the absence of a material witness for them who had broken his leg, and said they had been advised that this would be a sufficient ground for application for a postponement of the trial but that, conscious of their innocence, they were anxious at the earliest moment to meet the charge.

The case for the defence opened with the calling of a farmer named Simon Down to give evidence. He told the court that he lived at Durdon Farm, about 8 miles from North Tawton beyond Winkleigh Town. 'I knew the prisoner, John Taylor and I remember the 14th July last, Winkleigh Fair Day. Taylor came to my house about 9 o'clock in the morning of that day. He bought a horse off me and he stayed until half-past-eleven in the forenoon. His father was with him.'

Questioned by the judge, he answered: 'He bought the horse

for his brother-in-law and left it at my house. I sent the horse to Barnstaple. I have known the prisoner for ten or twelve years and I also known his wife. They carry packs of goods about the country and have honest and upright characters.'

The blacksmith of Winkleigh, John Leatheren, then gave evidence. He said: 'I saw John and Richard Taylor at Winkleigh Fair between two and three in the afternoon. I asked them to my house and they took tea there and stayed about an hour. We then went to Mr. Hammett's, the Seven Stars Inn, and stayed about half an hour and then went back to Walter Winn's, the Ring of Bells where we stayed until between seven and eight o'clock, when I left. I do not recollect whether I left alone or whether we all left together.'

Mary Bowden was then called. She said she was the wife of John Bowden and that they lived two miles from Winkleigh. She explained that there were only two houses where she lived and they stood together. 'I recollect the fire at North Tawton on July 14th. John Taylor and his father came to our house about 10 o'clock that night and asked to sleep the night in readiness to go to the races next day. They slept at our house that night but we did not immediately go up to bed. It might be half past ten or later before we did. We have only one upstairs room. They slept in the same room as we did. About 2 o'clock in the morning I was called up by my niece, Grace Quick, when she came home from the Fair. I got up to let her in. I spoke to John Taylor who asked Grace Quick if the town (Winkleigh) was pretty quiet now. She said it was not as "the travellers" (gypsies) were quarrelling. The clock hadn't struck six when I got up in the morning. We all got up about the same time and breakfasted together about eight o'clock. We had dinner about half past ten because they and my husband were going to the races and they went away together. We had not then heard about the fire over to North Tawton.'

Further questioned, Mary Bowden said they left about an hour before midday and went across the fields to the racecourse.

John Bowden, in his evidence, said: 'We never heard of the fire until we came to Winkleigh racecourse. Taylor said to me he was going to seek for some North Tawton folk to hear the truth of it and if it was true he would go home directly. He went off and

I didn't see him afterwards.' Grace Quick confirmed her aunt's testimony and William Gill, who lived in the house next to the Bowdens, told the court that he saw John and Richard Taylor go into the house adjoining his that night.

Loveday Hooper, daughter of the prisoners at the bar, said she had been in the churchyard on the night of the fire and that, had her father and grandfather been there she must have seen them. 'When my mother came to my house with our child she was in great distress.'

Questioned by the judge, she said that at her father's house on Monday things had been in their usual state. Her mother had been unwell and had taken medicine on the Monday and added: 'She used to sleep in a flannel petticoat when unwell. My husband has broken his leg and that is the reason why he is not here.'

The reference to the mother wearing a flannel petticoat when unwell is a reminder that the privy was at the bottom of the garden and, furthermore, that medicines were nearly always purgatives in those days.

The next witness was John Taylor's brother-in-law, Robert Bradford. 'I asked him to buy a horse for me and he did so from Mr. Down whose son delivered it to me.'

The defence then questioned the implication that Taylor had caused boxes of his goods to be removed from the house immediately before the fire (as suggested in Sybella Pike's evidence). The defence called Ann Churchwarden, who told the court that she was the daughter of Richard and sister of John Taylor who lived at Highweek near Newton. She said she had visited her family at North Tawton and had returned home on 12th May (1834). She said: 'I took away my box with me. It was mine before I was married.' Here the witness enumerated several other packets and articles which she took on returning home in a cart, such as a basket of things for a neighbour, a parcel for a man at Kingsteignton, a bag of potatoes for another and several little commissions which she had undertaken for her brother and others and she continued: 'I know Sybella Pike and saw her standing by our cart as I was about to move from my father and brother's door.'

Evidence was then taken from several drapers of Exeter from

whom Taylor had been in the habit of buying his stock. They all agreed that the prisoner had been honest and had paid for the goods he bought, spending, in all, about £100 a year.

This concluded the case for the defence and the judge gave his summing-up. He pointed out the various divisions into which the evidence fell, such as that which dealt with the question of John Taylor's whereabouts on the night of the fire, and that concerned with Jane Taylor's actions and remarks, the evidence of the insurance agents and the suggestion that the goods had been over-insured. On that point had not the insurers accepted payments from Taylor for two years?

After a short consultation the jury delivered its verdict of not guilty as to both prisoners, after which the learned judge said that the prosecution had been a necessary and very proper one and that the parish would have been wanting in its duty if they had not caused it to be brought. Turning to the jury he added: 'But, gentlemen, I quite concur with your verdict.' He said he thought Taylor had brought trouble upon himself by loose talk in public houses about dishonest acts with insurance agents.

The prisoners were then arraigned on another indictment charging them with the same offence as before with intent to defraud John Turner, their next-door neighbour, but no evidence being offered they were again acquitted and the business at the Crown Bar closed precisely at 4 o'clock.

The case must have been entered into without sufficient care being taken by the parish authorities to distinguish between gossip and fact and the behaviour of Jane Taylor in locking the door of her burning house would surely have suggested something underhanded going on, but the jury thought otherwise. When the houses were rebuilt did the Taylor family return to North Tawton? Did John and Richard collect the £150 from the Norwich Union?

The evidence given in this case sometimes seems to have little relevance to the charges against the prisoners but it reveals, in quite remarkable detail, the conditions in which the working people of Devon lived and thought at that time and throws a clear light on behaviour patterns and attitudes. Perhaps the most surprising aspect of rural life in the 1830s for the labouring class was the utter lack of personal privacy, as revealed in the evidence of those who told of the use by family and strangers of a single room in which all slept, and bedrooms without ceilings that were directly beneath the thatch. Nor is there any hint of resentment but rather an acceptance amounting to quiescence, as though such lack of privacy and such primitive conditions were, for them, right and proper and above all quite satisfactory.

Among the 20 indictments against the Taylors were some alleging that the burning of Taylor's house also did criminal injury to the man who leased the property and rented it to Taylor, to the man who actually owned the structure and to the ground landlord all of whom stood to lose by there being no rentable house left standing on the plot. Thus the network of interdependence between the various strata of social class is revealed. Even the lord of the manor was directly dependent on the tenants of those who leased his land for a part of his income and for many of the services he enjoyed. The labourer, in his turn, was equally dependent on the classes above him and, in a peculiar way perhaps, to the benefit of all.

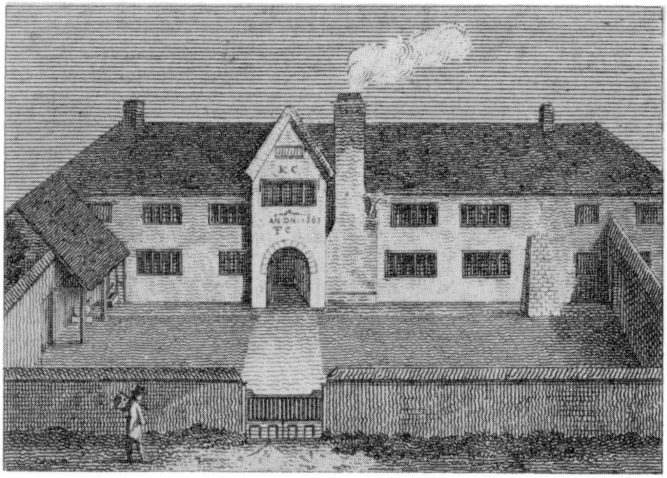

BARTON HOUSE, NORTH TAWTON, 1803

LYMPSTONE
FROM THE TERRACE

6

LYMPSTONE

There can be few prettier places in all Devonshire than the riverside village of Lympstone on the east bank of the Exe. It is a place of placid waters upon whose surface tall trees are mirrored and where, throughout the whole of the 19th century, a great fleet of fishing vessels of varying size and design sailed in and out of the estuary. Most of the craft were engaged in fishing but many of the larger ships were coastal traders which brought goods from Continental ports to Exeter or coal from Wales to feed factory furnaces of the ever-growing industries around the city.

The foreshore in the year 1833 was a busy place, with women tending the shellfish beds and some of the men engaged in net-mending or boat-repairing. The population of Lympstone at that date was probably less than 800 persons yet it seemed to enjoy a position of some importance, having no fewer than three postal deliveries daily from Exeter, the first at 5.29 a.m., the second at 10.33 a.m. and the last at 4.28 p.m. with outgoing mail leaving at 9.00 a.m. and 3.00 p.m.. The reason for so small a place receiving such attention may have been the number of important persons who resided in the many large houses in the vicinity, among whom were five churchmen, three Royal Navy captains as well as several high-ranking Army officers. Sir Trayton Drake occupied Nutwell Court on the edge of the village and a short distance away was the home of Sir George Burgmann. No doubt, too, some successful Exeter businessmen resided thereabouts, preferring the tranquility of the village to living in the noisy, smelly city.

On Wednesday, 12th June 1833, a lighterman named Lyddon got up from his bed as dawn was breaking and, as was his custom, kindled a fire with furze to cook his breakfast before going down to his boat and his day's work carrying cargo from ship to shore either down river to Exmouth or up river to Topsham. He left his cottage between 4 and 5 o'clock that morning, not knowing that some sparks from his fire had dropped from the chimney on to the thatched roof, where they lay smouldering for about an hour until the strengthening wind fanned them into flame. Lyddon's cottage stood among a cluster of similar dwellings near the water, between the bridge and the cliff.

The streets there were narrow and the houses packed close together, their straw thatched roofs tinder-dry from several weeks of warm summer weather so that the fire spread with alarming speed from roof to roof. Even at that early hour most of the men and many of the women were already at their work in the boats or at the mussel and cockle beds. Thus, when the fire was discovered there were few but the mothers and babies, the elderly, the sick and the children. At first there was considerable confusion, women and children ran from street to street, cottage to cottage to see if relatives or friends were aware of the danger and soon they were all seeking to get away from the heat and smoke of the blazing thatch which fell from the roofs, yet there seems to have been nobody capable of any positive action, and as a result valuable time was lost in summoning help.

When news of the fire reached the larger houses, the occupants hurried to the scene and directed the efforts to control the outbreak. Sir Trayton Drake sent in his fire-engine, a small machine of a kind often to be found at large country houses. Sir Trayton followed his engine and, with the naval captains and others, worked unceasingly to put out the blaze. The west wind was by this time so strong that burning debris was being scattered about in all directions even to a distance of 200 yards.

When some dozen cottages were well alight one of the gentlemen suggested that two or three in the path of the fire should be pulled down to confine it. There was evidently some disagreement on the point: perhaps the owner, or owners of the property in question were optimistic enough to hope the fire could be put out before demolition became necessary. Whatever argument prevailed we shall never know but certain it is that the flames swept on to engulf streets, courtyards and outbuildings, stables and pigsties.

An unaccountable delay seems to have occurred in sending for assistance to Lympstone's neighbours and two and a half hours elapsed before the Exmouth fire-engine arrived from a distance of barely two miles. A few minutes after the Exmouth machine came the one from Topsham and, there being no shortage of water at the scene, they set to work to wet every building threatened by the burning particles blowing in the wind.

It was 10 o'clock before news of the outbreak reached Exeter and the West of England Insurance Company's famous engine, The Little West, set off immediately, drawn by four horses hired from the livery stables at Bedford Street, and the machine covered the eight hilly miles in 50 minutes. It was followed into the burning village a quarter of an hour later by the fire-engines of the Sun and the Norwich Union.

All the firemen from the various brigades worked in harmony and in spite of expert skills gained by experience, it was not until midday that the fire was properly under control, by which time no fewer than 58 dwellings had been reduced to a heap of smoking, blackened ruins. At one time it was feared that a large house on the side of the cliff, the home of Captain Adney, would catch fire but the timely arrival of the Exeter engines prevented this though not before the adjacent shrubbery had been severely burnt.

The first task of those benevolent gentlemen, when their blackened faces had been washed and their burns dressed, was to survey the damage and count the number of homeless people. Although most of the cottages destroyed were probably of the one-room-up-and-one-down type, so common all over the county, many housed two families and the total number of the homeless was 240, an average only slightly more than 4 persons per house: a figure rather smaller than in many other towns round about.

Because of the rapid spread of the flames, little furniture had been saved by the inhabitants, some of whom had no more than a shift or nightgown and a blanket to cover them. The people who had come to their aid when the blaze had begun had spared no effort to save furniture from those cottages into which they were able to enter and some of it had doubtless been preserved, but the bedrooms, which were immediately under the thatched roofs and having no sort of intervening ceiling, could not be reached so that clothing and bed linen were the first things to be destroyed.

As soon as it was seen that the danger was over the people of the village gathered together to see what could be done for the sufferers. Those whose homes had escaped the fire offered shelter to as many as they reasonably could while barns and even linhays were prepared to shelter any who could find no better lodging. The gentry made generous donations towards the replacement of necessaries; the ladies gave clothing to the women and children and assisted in arrangements in preparations for feeding the homeless.

With their usual somewhat obsessive interest in the financial aspect of such disasters, the *Exeter Flying Post* and the *Exeter and Plymouth Gazette* both put the cost of the damage at around £6,000 but, unlike many similar occasions, they seem to have been quite unable to persuade anyone to disclose whose property had been insured and for how much, the local gentry evidently deciding it was nobody's business but their own.

In commenting on the fortunate fact that no life was lost in the fire, the *Gazette* draws attention to the sterling work done by Mr. Harrison of the Globe Inn at Topsham who was in charge of that town's fire-engine, by whose efforts, the newspaper asserts, much further damage to property was averted.

A general gathering of the inhabitants of Lympstone took place that afternoon and while the acrid stench of the burnt cottages still hung on the summer air, further donations were pledged and the rebuilding of the cottages discussed as a most urgent matter. The most notable outcome of the fire was that it changed the appearance of that part of Lympstone by a unanimous determination to abandon thatch for roofing and, further, it was decided not to reconstruct cottages with the highly combustible materials used in cob walls. The new cottages which replaced those burnt that day were almost certainly larger, despite being erected in the same narrow streets, because portions of land that had been back yards were incorporated. They were built of brick and had slated roofs and it is not difficult to imagine the bustling activity in the village during the months of rebuilding but it is less easy to guess what were the feelings of the sufferers immediately after the disaster as they stood before the doused, stinking ruins of what had been their homes. Among the cinders must have been recognizable fragments of cherished possessions, shattered china and glassware, perhaps the collection of a lifetime, and the lovely early summer day must have seemed to mock their depressed spirits. Was their attitude towards their possessions less materialistic than is said to be the case today? Such things were probably just as cherished but with rather more sentimental feelings. The contents made the 'home' rather than the house itself and to a degree that could inspire the ballad, 'Home Sweet Home', a sentiment widely shared during the 19th century if the great novelists of the period are to be believed. This attitude may account for the evident harmonious relationship between the gentry of Lympstone and their poorer neighbours and may also account for their spontaneous acts of practical help and sympathy. Among the buildings gutted was the village's poor house, whose inmates helped to swell the number of those completely destitute to 96 out of the 248 persons made homeless.

The *Western Luminary* reported that: 'Immediately after the principle danger had subsided the gentlemen of the Parish met at the Church, The Rev. Mr. Hull, the Curate, acting as chairman and entered into a liberal subscription on the spot amounting to £120 . . .' It is regrettable, however, that none of these leisured and educated gentry left behind them an eye-witness's description of the catastrophe in journal, diary or letter which might have given posterity a reliable and detailed account of the behaviour of those present: a record of one of those rare moments when the strengths as well as the weaknesses of suffering humanity are exposed to the observer.

7

MORETONHAMPSTEAD

Until the railways came to Devon about the mid 1840s communications between the towns and villages depended entirely on someone going from place to place either on foot or horseback or by horsedrawn wheeled vehicle. The railways were the fastest known means of travel and soon after the steam train made its appearance, the electric telegraph system of communication was introduced between railway stations and signal boxes throughout the entire length of each rail route so that a message, in Morse code, could be sent from, say, London all the way to Bristol almost as quickly as it could be tapped out on the telegraph machine. This innovation, coming so soon after the introduction of the steam engine and the laying of railway lines all over the country, brought about many drastic changes in social life. The very fact that information could pass from one end of the country to the other in a matter of minutes instead of several days introduced a new dimension into life as its influence upon business, politics and social intercommunication developed along the humming wires that were stretched from post to post all along the side of the railway tracks.

The small market town of Moretonhampstead suffered a number of damaging fires during the 19th century, some of the worst before the advantage of telegraphic communication could be used to summon help quickly. Quite early in the century the prudent inhabitants had provided themselves with a large fire-engine and, except during periods of drought, there was a fair supply of water readily available. In the year 1816, for example, there occurred two bad fires. The town had, at the time, a population of about 1,800 and when fire broke out there was never a scarcity of men to manhandle the fire-engine to the scene of the blaze nor to pump, or fetch and carry buckets of water to fill the engine's tank but, if the one engine was not able to contain the fire and help was required an express messenger was despatched to summon aid.

Moretonhampstead stands to the east of Dartmoor and is almost surrounded by hills. It is 12 miles west-south-west of Exeter, 12 miles north-north-east of Newton Abbot and 12 miles north of Ashburton and it was natural, therefore, when in need of assistance that the inhabitants turned to Exeter, the county town. Because of the distance, however, and the time it would take to send for help and for it to reach the town, the inhabitants made a practice of turning out quickly, and in force, whenever the cry of 'Fire!' was raised and they generally managed to deal with the outbreak themselves. On the 13th January, 1816, for example, the townspeople successfully dealt with a fire which broke out near the centre of the town. Several houses were quickly ablaze and a stiff breeze spread flaming particles of thatch from house to house and street to street so that at one time it was feared the entire town might be destroyed. As was so often the case in these small-town fires, the blaze started in one of a group of labourers' cottages and spread rapidly to a number of workshops which, together with out-buildings, stables, etc., were destroyed. The fire burned for some four hours and left fifteen dwelling-houses in ruins but due to vigorous use of the fire-engine and the deployment of men with buckets of water, who got on to roofs of nearby houses, the fire was confined and eventually put out. Nevertheless, the occupant of one small cottage, an elderly woman, was overcome by the smoke when her house caught fire and she perished in the flames before anyone could reach her.

On Sunday, 24th November of the same year at an hour after midnight another fire started in a court off a street leading to a paper-mill. Again there was a strong wind blowing from the north-east and, once more, the town was in imminent danger of being destroyed but, by the great exertions of the inhabitants employing the same methods they had used in January, that is to say sending men on to the roofs of surrounding houses, and keeping them well supplied with buckets of water to wet the thatch, while the engine pumped water into the blazing buildings, the fire was contained. On this occasion only three dwelling- houses were burnt, together with the Wesleyan Meeting House, two barns, a stable and several small workshops used by local craftsmen.

On 11th September in the year 1839 another, and more devastating fire broke out in Moretonhampstead and again it was discovered about 1 o'clock in the morning. In the issue of 13th September the *Exeter Flying Post* reported the event in detail, commencing thus:

> It is with deep regret that we announce the occurrence of a most destructive fire in the town of Moretonhampstead. About 1 o'clock on Tuesday morning the inhabitants of the town were aroused from their slumber by the appalling cry of 'Fire!', which was succeeded by the violent ringing of the alarm bell and so great a light was thrown over the whole town as to give but all too convincing evidence of the rapid progress of an awful conflagration.

The fire started at the White Horse Inn, kept by Samuel Gray, and spread quickly to buildings to right and left. A high wind, generally from the north-east but gusting in all directions, scattered burning fragments of straw so that in a few minutes the many outbuildings behind the White Horse were blazing, their thatched roofs being dry from lack of rain.

Lodged in one of these outbuildings, the slaughterhouse, was a small barrel of gunpowder which a farmer was to have collected a few days later. So rapidly had the flames spread in the wind, and because of the early hour, it was some ten minutes before the men were on the scene. Samuel Gray immediately warned them of the gunpowder and they kept well away until it had blown up with a concussion that shattered windows at some considerable distance around the inn.

Adjoining the White Horse was a substantial house in which lived a lady of modest means named Mrs. Stevens and next door to her on the other side from the White Horse was the town's most important staging-house, the White Hart Inn. Mrs. Stevens' house and the White Horse were doomed and blazed fiercely, sending red-hot sparks high into the air to be caught by the wind and scattered on the thatched cottages in Pound Street and Court Street. As soon as it was realized that it was impossible to save Mrs. Stevens' house the firemen concentrated their attention on the White Hart, clambering on to the roof in order to pour buckets of water over the dry straw to stop it igniting.

It quickly became necessary, however, to take the engine into Pound Street, where houses on both sides were ablaze. At this point it was already apparent that help was required from outside the town and a messenger was sent on horseback post haste to Exeter where he arrived soon after two-thirty.

The express rode straight to the police station in the Guildhall where the duty officer immediately sent all his men to alert the brigades of the various insurance offices, who were soon ready to leave the city.

The West of England's engine, The Little West, fully manned and with all apparatus and drawn by four horses, set off for Moretonhampstead. Shortly after it had left, the Norwich Union engine, also fully manned and equipped, followed. A moment later the Sun engine was also on the road.

The operation had been set on foot very quickly in the prevailing circumstances. The night police had to run to each fireman's house and waken him, and as each fire-engine had a complement of six men, the police would have had to find at least eighteen men, probably more, who would have lived in different parts of the city. The fire insurance offices did not have their own horses and the teams would have had to be made ready and harnessed to the fire-engines, which would have been manhandled to the stables by the firemen, a fairly time-consuming process.

The West of England engine was the first to go, at a little after 3 o'clock, having got its horses from Mr. Cochran of the New London Inn, but its luck was out, the traces parting on two occasions during the 12 mile journey.

As it is so many years since the harnessing of horses to carts and carriages was widely practised and generally understood, it may be appropriate to describe the kind of harness which would have been used by the fire-engines. The front wheels of the engine were pivotted on the axle so that the horses were used to steer the vehicle. The steering gear was in the form of a frame with a way-bar which controlled the radius and limit of turning while the horses were attached by the traces to another called a splinter-bar. These were fastened to the splinter-bar by four large studs called roller-bolts. From the centre of the frame was a long pole with a steel hook at either end and two pole chains, each with a small hook on the end. The wheel-horses (wheelers) were stationed on either side of the central pole and fastened to the vehicle by the traces which were made of double strips of stout leather which extended from the roller bolts on the splinter-bar to fastenings about half-way down the horses' collars and passing through the saddle tabs which, like riding saddles, the horses wore on their backs with girths to keep them in place. Being such an important part of the equipment the traces were also secured at these points by fastenings called tugs and the pole chains were hooked into the collars. The traces of the leading two horses were hooked onto a wooden bar, the leading-bar, which had an eye on its far side to hook into a longer one of similar kind, the main, or long-bar. The outer reins held by the driver reached the leaders through terrets, or eyelets in the saddle-pads of the wheelers then on through more terrets to the upper part of the head bridles. From here the reins continued on through the terrets on the saddle-pads of the leaders and were fastened to the operational ends of the leader's bits. There were also coupling reins between each pair of horses to prevent their heads turning in different directions. When all this harnessing had been done the driver had to run through an overall routine check of all the fastenings, buckles, hooks and to inspect the leather traces and reins to make sure everything was in perfect order. A broken trace at a full gallop could mean disaster both to the horses and the firemen. This method of harnessing also applied to the coaches and, so skilful were the men who were responsible for changing teams at the coaching-inns, that it was not unknown, when the mail coach was running late, for the complete change of horses to be made in one and a half minutes. When a trace broke, the safety of the horses, people and vehicle depended entirely on the skill of the man at the reins. In the prevailing circumstances it was quite remarkable that The Little West reached Moretonhampstead soon after 5 o'clock that morning, having twice had traces break, and after covering twelve strenuous miles of hilly roads.

The Norwich Union engine hired horses from Mrs. Street, of

the Royal Clarence Hotel, and left Exeter some twenty minutes after the West of England. The Sun Fire Engine left soon after a quarter to four and was drawn by four horses from George Pearse's livery stables in Bedford Street and it made the journey in the quickest time of one hour and ten minutes which, when its weight is considered, not only of the equipment but of the six men it carried, was a remarkable feat. Despite the broken traces The Little West suffered she still arrived first.

Samuel Gray, of the White Horse, was the greatest sufferer in this fire for not only was his inn practically gutted, but the many outbuildings behind were either blown up or burnt down, together with the entire contents of each which included eleven fine pigs just coming to maturity. Nine other houses were also gutted but at the White Hart, thanks largely to the indefatigable efforts of the firemen, who were helped by the innkeeper's neighbours, the damage suffered was slight and the business of the inn was carried on without interruption.

In the *Exeter and Plymouth Gazette* of 15th September a report on the fire states that, in Pound Street '. . . the fronts of nine or ten houses exhibited at the same instant one complete mass of fire.' If the fronts of the houses were alight it suggests that they were probably constructed entirely of timber. The report also states that the spread of the blaze was stopped by a house with a slated roof and which, it may be supposed, was built of brick or stone. Court Street suffered less damage because, being at a little distance from the White Horse, where the fire started, men had time to get on to the roofs with relays of people supplying them with buckets of water. Despite this, however, a number of workshops, pigsties and linhays were burnt down in the immediate vicinity. Fortunately no lives were lost on this occasion but fourteen poor families were left homeless.

By the year 1845 the woollen industry had so far declined in Devonshire as to be little more than a few looms worked principally for the purpose of supplying local requirements and the inhabitants of the small towns seem to have settled for a very modest living derived from providing services and goods for local consumption. Thus, in a report of a dreadful fire which broke out at 1.00 a.m. on 12th September 1845 the *Exeter Flying Post* said: 'We deeply regret to state that a fire, most destructive in its nature and calamitous in its result occurred in the town of Moretonhampstead in this County on Friday last when 40 houses were destroyed and about 50 families deprived of their habitations.'

The blaze was discovered by a man who on happening to wake saw flickering lights through the window of his bedroom. By the time he was able to alert neighbours, the fire had a firm hold on a number of houses. People at that time were well aware that speed in getting out of the house was essential and on this occasion no life was lost and the newspaper report continues: 'the fire broke out in the bakehouse of Mr John Dayment in Cross Street . . . . The bakehouse was to the rear of the premises near Back-lane.' The fire caught light to 'The Workshop of Mr. Simon Bellamy, of the Golden Lion Inn who is also a wheelwright.' Evidently those who were able carried on more than one trade to earn their living.

Most of the houses were, of course, covered with the usual straw thatch and the fire spread quickly both to right and left of the bakery and rapidly ignited houses and outbuildings to north and south. Running westward it fired the back premises, including the stables, of the Golden Lion, the premises of Simon Neck the draper, and on to the old schoolhouse. Nor was the blaze confined to these back lanes but sparks were blown on to the backs of the houses in Fore Street which instantly became a mass of flames. An unusual feature of this fire was the nature of some of the buildings ruined. Whereas most dwellings were of cob construction some were massively constructed of blocks of cut granite. Such, among others, was the old schoolhouse but so fierce was the fire that it was reduced to a heap of ruins. Another building to perish was known as 'Mount Arthur', the facade of which as the newspaper put it '. . . bore testimony to its antiquity with its once famed marble hall.'

At the time of this fire Moretonhampstead boasted two good fire-engines and they were placed in what was considered the best positions to do greatest service. There was no shortage of manpower to pump and to fetch and carry buckets as it was the invariable practice, on such occasions, for every able-bodied man to go to the scene as fast as he could. Many of the women also helped to turn well-handles and fill buckets.

A messenger was despatched to Exeter for assistance but, by arranging human chains to pass buckets to and from neighbouring brooks the local engines were well supplied with water.

The battle continued throughout the night and the efforts of the inhabitants confined the blaze to the one area. Fortunately, too, the north-easterly wind, which was blowing hard when the fire started, decreased and the hazard of flaming straw being blown about was reduced to manageable proportions.

The church and the churchyard became repositories for piles of furniture but the owners seem to have taken care that none of it was smouldering when taken to safety. At the height of the blaze, the strong wind carried the smoke away southward and the flames lit up the face of the church clock so that its hands could be clearly seen at a great distance.

The assistance requested from Exeter arrived at about eight o'clock by which time the fire had been extinguished. The *Exeter Flying Post* report contains the names and occupations of some of the people who lost their homes and goods and is included here to show by what means they earned their livings, bearing in mind that the population of the town was under 2,000. The railway was being built but had not yet had the devastating effect it was destined to have on the 'passing trade' of the coaching-inns.

The report states: 'Among the sufferers on this most distressing occasion may be mentioned Thomas Holman out-pensioner of Greenwich Hospital, Richard Taylor carpenter, Francis Williams granite mason, J. Symons labourer, William England excavator, William Taverner carpenter, John Hill blacksmith, Alexander Hutchings thatcher, Mr. Hill schoolmaster, Joshua Horrell cordwainer, James Salter small farmer and retailer of cyder, William Walling carpenter, T. Croot sawyer, S. Bellamy victualler and wheelwright, S. Neck draper, Henry Endercott farmer and butcher, J. Eaver granite mason, J. Dayment baker, W. Howe cordwainer, W. Trace small farmer and dealer in timber, Henry Short labourer, etc. etc.' These are representative of the fifty families rendered homeless, of whom, the newspaper told its readers, many were uninsured and a subscription was started for their relief. In the same paper's issue of 16th October 1845 (almost exactly one month after the fire) the following appears: 'The Committee appointed to consider the best mode of alleviating the distress occasioned by the late fire in this Parish beg to inform the public, and those who have so kindly contributed funds for this purpose, that enough has already been received to indemnify the sufferers.' Moretonhampstead, it would appear, had a better fund-raiser than any of the other small Devon towns which suffered from disastrous fires and the speed with which the target was reached must have been a record.

The total commitment of all the inhabitants to give help in the event of fire breaking out in their town suggests an intimately integrated society in which the art of living by supplying one anothers' requirements could be practised successfully even if, as was probably the case, it produced the narrowest of parochial outlooks so that each small town became a kind of island visited by travellers who were likely to stay for only a few hours at most. The coaching-inns would have been the centres of news and gossip about the world beyond the narrow confines of the town. When the railways came the change in communications must have been such as was bound to gradually break up the isolated intimacy of these small pockets of society as it widened the horizons of the active young and made visiting other places a simple and, from the point of view of time taken, a practical proposition for everyone.

Nine years after the 1845 fire another similar disaster struck Moretonhampstead at about two o'clock on the morning of 10th August 1854. Like the previous outbreak it was discovered accidentally when a man was awakened by the smell of burning and on opening his eyes saw the reflection of flickering flames on his bedroom wall. Pausing only to put on a shirt and trousers before arousing his neighbours he discovered that the fire already had a good hold on a number of nearby dwellings and outbuildings. The fire was in streets of modest to poor houses, built close together and all covered with thatch as, too, were the stables, pigsties, workshops and other out-buildings. These quickly caught fire as the wind blew flaming straw from roof to roof and eleven houses in Court Street were destroyed and in Pound Street, which had suffered badly in the blaze of 1839, ten more were burnt down.

The fire, it appeared, had started in a stable belonging to a Mr. Weeks and, while some men ran to get the town's two fire-engines others hurried to extricate the terrified horses and to let loose the pigs and poultry. The horses in the main stable building were successfully taken out to the safety of a nearby field but Mr. Weeks' own saddle horse, which was in a separate building, perished. In the same courtyard, between Court Street and Pound Street, were some workshops and a carpenter lost all his tools and a quantity of mature timber as well as his place of work.

The people of Pound Street had not, however, forgotten the fire-fighting strategy adopted so successfully in the 1845 conflagration and quickly organized themselves accordingly. Everyone possessed of a bucket ran with it to the scene of the outbreak and ladders were brought so that men could get on to the roofs of houses not yet alight, there to brush off any burning embers blown by the wind and when necessary to pour water on the thatch to prevent ignition. The two local engines were kept supplied with water by human chains passing buckets filled at the nearby brooks. One engine was placed in the middle of Court Street, the other in Pound Street at no little hazard to the men operating the long-handled pumps because the streets were so narrow and the danger of blazing thatch falling on them was ever present. This, however, meant that only the dwelling-houses could get the benefit of the engines while the stable and other buildings in the curtilage between had to be dealt with by throwing buckets of water on them. Such were the efforts made by the many people attending the blaze that the spread of the flames even in so congested an area was confined to a comparatively small amount of property. There was also an element of good luck in this as the wind dropped almost as soon as the fire was discovered, and after strenuous efforts by the inhabitants the fire was first confined then put out with no mishap to anyone, although some twenty families were rendered homeless.

During the struggle to extricate Mr. Weeks' horses questions were being asked as to how the fire began and dark hints were bandied about that it had been started deliberately. The rumour spread and innuendos flew thick and fast while those deprived of their homes naturally grew more and more angry towards the unknown malefactor and the mayor decided it would be prudent to order an enquiry. On the Tuesday following the fire, therefore, an inquest was held at the White Hart Inn by the coroner, W. A. Cockey, Esq, but so many crowded into the inn and so few had any relevant information to offer that, after making every effort to discover a witness to the act of arson, he adjourned the inquest for fourteen days, by which time sober reflection seems to have made the rumour-mongers take a more rational attitude. The resumed inquest fared no better than the first and fewer people attended. Mr. Weeks was probably questioned as well as the man who gave the alarm but, as both had been abed when it started, there could have been little for them to tell and the matter was indefinitely adjourned and eventually forgotten. The houses would have soon been rebuilt or repaired, and the subscription which had been instantly instituted adequately compensated the sufferers so that life in Court Street and Pound Street, Moretonhampstead must have returned to normal in a very short time.

MORETONHAMPSTEAD
by John Gendall (*c* 1830)

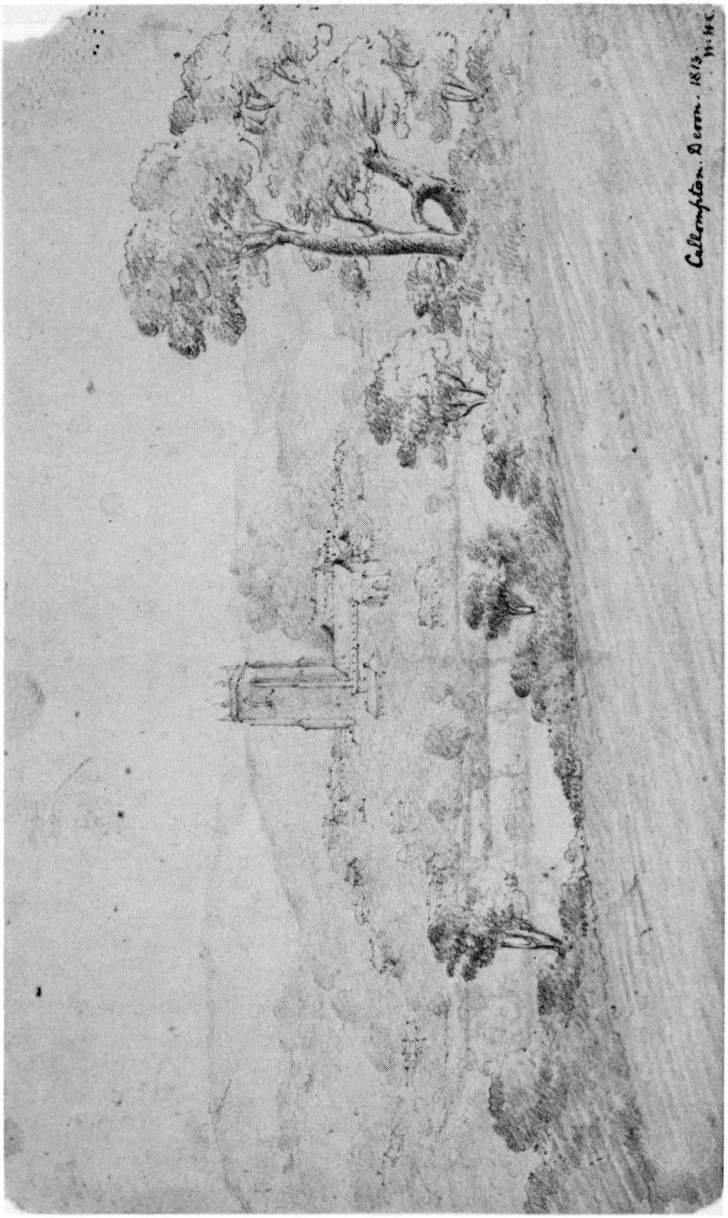
Cullompton. Devon. 1813.

8

CULLOMPTON

The country dweller of the first half of the 19th century appears, from the evidence of written and printed records, to have been steeped in the traditions of his locality to a point where even the most rational demand for change was either directly opposed or simply ignored. Because of this stubborn attitude decades were to pass before the highly dangerous straw thatch was, as a dwelling-house roofing material, abandoned, although its use on outbuildings continues in Devon to the present day.

The prolongation of this use for straw may, to some extent, have been due to the mushroom growth of fire insurance companies who, for a few shillings a year, would indemnify the cost of rebuilding any property destroyed by fire. The average two-roomed cottage, such as was occupied by a labourer's family, could be built on traditional lines for less than £20 and it was arguable, therefore, that the cost of such a building, plus the cost of insurance, was still considerably less than for the more fire-proof brick-and-slate structures. Thus, economic considerations may have been as much responsible for the reluctance to change as was tradition, while the combination of the two incentives would have been practically irresistible.

The small town of Cullompton was a typically traditional Devonshire market town comprising a long main street with buildings on either side and intersected by short side-turnings with their rows of small cottages, behind which were stables, workshops, linhays, pigsties, etc. and almost all, like the dwellings, were roofed with straw and thatch, safe enough in wet weather but, in times of drought, a fearful fire-risk.

Fires occurred with alarming frequency in every decade from the beginning of the century, until the 1870s at last brought uniform change from thatched roofs to slates and in many cases from cob walls to brick.

There was a serious fire which destroyed a number of houses in March, 1805, and on Sunday, 29th June 1806, at about one thirty in the morning, the premises of Mr. Hall, the baker, were found to be on fire. The flames had got a good hold on the bakehouse before the danger was discovered and nearby buildings were soon also alight. It took the inhabitants several hours to completely douse the blaze, by which time the bakery and five dwelling-houses had been gutted. So rapidly did the fire spread and so furiously did it burn that almost no furniture from the five houses could be saved. Several other nearby cottages were also damaged but their contents were salvaged, doubtless because even at this early date Cullompton was able to call upon the services of two local fire-engines which, although unable to quench the fire in the blazing buildings, were nevertheless instrumental in preventing it spreading beyond the immediate neighbourhood.

By eight o'clock that morning the efforts of the firemen and the many willing helpers succeeded in putting out the blaze but not before half a dozen families were homeless. Although nobody had suffered injury or death, several people had been forced to leave their homes by way of upper windows on shaky ladders 'borrowed' from wherever they could be found.

Less than one month later, on 25th July, what might have been a serious fire was averted at the toll-house at the Western Turnpike by the timely arrival of the detachment of the Worcester Regiment of Militia who had marched into the town less than two hours before the outbreak. Newspaper reports of this period give several instances of valuable assistance being rendered by soldiers who happened to be in the vicinity when fire broke out and their disciplined training often turned chaos into order and reduced time-wasting in procuring buckets of water to a minimum.

In its issue of 2nd September 1815, *Woolmer's Exeter and Plymouth Gazette* began a report of a fire thus: 'Cullompton, so often ravaged by fire has again been the victim of its scourge' and went on to tell another story of an accidental fire which happened in the workshop of Mr. Frost, a cabinet-maker. At nine o'clock on the morning of 29th August, some of his workmen went to breakfast leaving the glue-pot over a fire upon which they had piled a quantity of wood shavings. On their return the workshop was a mass of flames of such intense heat that their attempts to put it out, by throwing buckets of water on it, were fruitless. Particles of burning thatch and shavings were carried into the air to be blown by the breeze on to nearby houses which caught fire and blazed like torches. All the efforts of the inhabitants and of the firemen were at first incapable of getting the fire under control until several buildings had been stripped of their thatched roofs to impede the spread.

When at length the fire had burnt itself out twenty-one houses and about half that number of outbuildings had been reduced to ruins. Once more, by getting out of their burning homes quickly and without attempting to save more than they could put in pocket or upturned apron, all the occupants escaped with their lives.

In view of the ever-present danger of fire starting in such inflammable buildings, it is surely astonishing to learn of such carelessness as shown by the workmen employed by poor Mr. Frost. They must have been perfectly well aware how easily glue catches fire when it boils over: these men seem to have taken no precautions against such an occurrence and upon discovering the disaster probably put the blame on the apprentice or the cat.

Looking back on the many fires which occurred at Cullompton in the early part of the 19th century it is surprising how few practical precautions were taken either by individual residents or by town council for the protection of property from fire. The outbreak of the 29th August 1815 demonstrated the carelessness of workmen, possibly only excusable on the grounds of simple thoughtlessness but no excuse can surely be offered for the local authority's criminal neglect of the town's fire-engines when, on Wednesday, 22nd June 1831, at about midday, fire broke out on the premises of Charles Selwood, a tanner and maltster. *Woolmer's Exeter and Plymouth Gazette* remarked: 'It most unfortunately happens to this town that whenever a fire does take place that its ravages are most destructive. The two Fire Engines were out of repair on the present, as on former occasions.'

In the circumstances the town council might have been expected to have appointed some person to maintain the two engines in a state of readiness and, as later reports were to show, the matter was put right soon after the above remarks appeared in several of the Exeter newspapers.

On this occasion the fire resulted from woodwork close to a flue catching fire when it became overheated. The blaze spread with remarkable rapidity and the fire-engines were soon dragged from their sheds but, to quote the *Exeter Flying Post*, they were both '. . . in such need of repair as to be for all effective purposes, useless'. The flames, therefore, had time to spread unchecked while people hurried here and there in search of buckets before the laborious task of throwing water by hand on the blazing houses could begin. It was quickly realized that the property in the immediate neighbourhood of the outbreak could not be saved and all efforts were directed towards preventing the fire spreading. Thatch was torn down from roofs and a lot of otherwise undamaged cottages received an unwelcome soaking. Twenty-one houses and tenements were gutted before the fire was put out while the great majority of the tenants, being of the poorer class and uninsured, lost much of their furniture and clothing. Among the wealthier losers was a John Rowe, a grazier, whose insurance cover had expired but a few days before.

Considerable assistance in preventing the spread of the fire was given by a fire-engine and brigade sent from neighbouring Tiverton, where the record of frequent fires must have given the men abundant experience.

The labouring and poor people were not always the sufferers from fires, as was demonstrated by an outbreak in Cullompton on Saturday, February 25th, of the following year when the house of Mr. Whitter, a banker, was discovered to be on fire. The blaze was so furious that in little more than an hour the building was reduced to a heap of smouldering ruins, as was also the bank next door, as well as four other fairly large houses adjoining. Everything in Mr. Whitter's house was completely destroyed while all hands worked with great zeal to save the contents of the bank of Hole, Whitter, Brown & Co., from which everything was safely removed before the building was also burnt down. Nor did the worthy bankers waste any time bemoaning their misfortune but took over the occupation of an empty house across the street and were open for business the following morning. All the burnt-out houses, it seems, were adequately covered by insurance as were the contents of each. Such precautions at least meant that houses and furniture could be replaced with no loss to the owners but it is open to speculation as to what unique treasures were constantly being destroyed in fires entirely due to the practice of roofing with straw thatch.

Under a heading dated London, 20th June 1837, the *Exeter Flying Post* solemnly announced, between broad black edgings, the death of His Majesty King William IV, and for the next three weeks all the Exeter newspapers bore the sombre signs of mourning. In its issue of July 1st, *Woolmer's Exeter and Plymouth Gazette* reported in an obscure column on a middle page the sad but brief story of another fire at Cullompton. It read: 'On Sunday last, the 25th ultimo, 9 houses were burnt down in Cullompton for which act a woman named Lock is charged and has been committed to the Devon County Gaol to take her trial at the Assizes.' The report is tantalizingly lacking in information and must have left readers wanting to know more. It is not unreasonable to suppose that some tragic human story lay behind such a desperate act because everyone knew that the punishment for setting fire to property was death, and during the previous year an Ottery St. Mary man had gone to the gallows for just such a crime.

Perhaps, however, the report may have been overlooked by most readers who would have been more interested in the news about the new Queen, the young Victoria, about whom they can have known little.

The *Exeter Flying Post* reported: 'Her Majesty held a Privy Council in the forenoon at Kensington Palace which was attended by the Archbishop of Canterbury, Viscount Melbourne and the rest of the Cabinet, the Duke of Wellington and a large number of Privy Councillors without any reference to politics as usual. The distinguished manner in which her Gracious Majesty acquitted herself astonished all present, not even excepting the Duke of Wellington.'

As if that was not enough to whet the appetites of readers, the paper continued: 'In grace and youthful dignity, as well as majestic composure, she showed herself every inch a Queen.'

No wonder the woman named Lock was overlooked, if not altogether forgotten as were, it would seem, the unfortunate occupants of the nine dwellings she was alleged to have caused to be gutted. If she was permitted to read the account of her arrest in the papers their black borders must have seemed an ominous sign to the poor woman, who the report had dismissed without even giving her the benefit of a Christian name.

Under the heading 'City of Exeter', the *Flying Post* went on to announce: 'Guildhall, 5th July, 1837. On Saturday next at 7.00 p.m. being time appointed for the interment of His Late Most Gracious Majesty, the Mayor trusts that the inhabitants of this City will manifest their respect for the memory of their lamented Sovereign by partially closing their shops until 5 o'clock p.m. and entirely closing them and suspending all business from that hour for the remainder of the day.' Quite how a shop can be partially closed is not revealed.

In the issue of the same newspaper dated 27th July, or about four weeks after the arrest of the woman named Lock at Cullompton on a charge of arson, appears under the heading 'Devon and Exeter Assizes' this brief announcement: 'Sarah Lock (at last given a Christian name) aged 46 was indicted for having, on the night of Saturday, June 24th last, at Cullompton, viciously set fire to a dwelling house, the property of Joseph Frost, and acquitted.' One may speculate on the woman's feelings after having spent a month in Exeter Gaol before being brought to trial—only to be found 'not guilty'. Was she arrested on nothing more substantial than malicious gossip or was the evidence against her so badly presented as not to satisfy the jury? We shall never know and poor Sarah's brief moment of notoriety ends in oblivion as complete as if she had never existed. Our curiosity as to what became of her is left unsatisfied but it also suggests that perhaps judges conducted their cases with rather greater care than they have since been given credit for.

Two years after the young Queen Victoria's accession to the throne, on a bright but windy Sunday morning, of 7th July 1839, the congregation was shuffling quietly from the parish church, the ladies holding down their skirts with one hand and grabbing their hats with the other as they gossiped with one another about domestic matters. The sun appeared in fitful flashes as the clouds scudded across the sky like patches of scum on a swift-running stream. As they stood for a moment after the morning service on the 'day of rest for all but Mother', someone observed that a chimney behind New Street, down at the lower end of the town, was belching black smoke. A moment later many saw a tongue of flame leap up followed by a shower of sparks which fell upon the thatched roof of a cottage and set it ablaze.

It was some minutes before anyone in the vicinity of the trouble was aware that anything untoward had occurred. By the time the flames had been seen by an elderly woman, who raised the alarm, a gust of wind had carried burning particles of thatch on to the roofs of a row of dwellings in New Street itself. In an instant these were blazing furiously and, in their turn, providing the boisterous wind with more burning fragments, which rose high in the air to fall on houses in streets at a great distance from the original outbreak. Within a quarter of an hour some six separate fires were burning, each one threatening to fuel as many more. Some of the congregation hurried home only to find their own houses already well alight.

In a very short time the whole town was in a state of alarm and the two fire-engines were quickly dragged to the place where it all started and were enthusiastically brought into action and were, this time, found to be in good order. The inhabitants of the immediate neighbourhood exerted every effort to keep the tanks filled with water but the fire continued to spread and an express messenger was despatched to Exeter for assistance.

Houses on either side of the Half Moon Inn, a full quarter-mile from where the blaze first started, were seen to be alight. This caused considerable consternation as the property there and the Red Lion were by far the most valuable in the town. In every part of Cullompton men had to scramble on to the roofs of their dwellings either to douse the already burning thatch or to wet it to prevent it taking fire and this, it would seem, they did with some success at first. The gusting wind, however, and in some places the difficulty in obtaining a constant supply of water from the street's leats, defeated their efforts.

In about two hours the help arrived from Exeter in the person of Captain Cummings on The Little West and followed soon after by the West Middlesex Company's Engine and shortly after by the Sun's brigade.

When the messenger had alerted the fire brigades the three who were to attend the blaze sent men to the stables for horses. None of the insurance companies kept stables of their own. A groom leading two of the horses for the West of England was waylaid by men of the West Middlesex who took them for their own use, having obtained only one pair from their usual supplier. This delayed The Little West for more than a quarter of an hour but Captain Cummings and his men eventually set off in pursuit of their rivals. Going down the road at a full gallop with whip cracking and post-horn blaring, The Little West fairly flew through hamlet and village until, six miles out of Exeter, it caught the West Middlesex, passed it with a hearty cheer, in a cloud of choking dust and, as was its wont, was first to reach the outbreak.

Immediately upon its arrival the Captain manoeuvred his machine into position to prevent the fire reaching the property on the green, and the task proved almost impossible, so strong was the wind and so many the roofs ablaze everywhere.

As soon as the West Middlesex engine reached the town, efforts were concentrated on trying to save the houses between the Half Moon and the Red Lion but the former inn was, by then, a blazing ruin.

Meanwhile the town's own engines were working in the narrow side-streets with dwellings and shops on fire all around them and the intense heat threatened to drive the firemen back and burn their engines, but with considerable courage they kept their station and slowly began to get the upper hand.

The Sun engine, too, rendered invaluable service that windy Sunday afternoon at another part of the town, where many houses were gutted. Horses, pigs and poultry had to be rescued and it must have been a remarkably lucky family that was able to sit down to lunch that day.

The Half Moon and most of the large houses in its vicinity were destroyed completely despite the valiant efforts of the West Middlesex Brigade. The usual story of water-shortage, so often met with in other places when visited by fire, was not the case at Cullompton on this occasion in spite of its being near midsummer. Had there been a drought the devastation must have been dreadful as more than one-hundred houses were gutted, and between forty and fifty stables, linhays, workshops and other outbuildings were destroyed. In addition there were more than fifty other dwellings partially burnt; many were beyond repair and had to be entirely rebuilt. So fierce had been the flames and so rapid their spread that little furniture was saved and the entire stock of many a shop was completely consumed.

The *Western Luminary* published the day after the fire reported: 'It was not until a late hour last night that the fire was extinguished and the engines played on the smoking ruins until 5 o'clock this morning.' The report went on to say: 'As may be supposed the town of Cullompton presents a melancholy appearance after this really awful fire. Whole streets have been burnt down, in fact two-thirds of the town may be said to have been destroyed.'

There was only one serious accident to an inhabitant when a partition wall collapsed and fell on a chimney-sweep, severely injuring his back, and he did not live to see the dawn.

The bank, which a few years before had been burnt down, escaped undamaged on this occasion although the adjoining houses were gutted again. After the previous blaze they had been rebuilt and all but the premises of the bank had been rethatched. Slates had been preferred by the bank and doubtless accounted for its escape.

Speculating on the cost of the damage caused by the blaze the *Exeter and Plymouth Gazette* considered that: 'Not only the buildings but also the greater part of the furniture and stock-in-trade therein must have been burnt and scarcely £100 for each building would be sufficient to cover it, and we think could make a total loss of above £10,000. We fear however that it will be found to be much greater.'

The immediate difficulty was to find shelter for more than one hundred and fifty families who had lost everything they possessed except the clothes they wore and who had nothing but their bare hands to earn their living with. So far as finding employment in the near future was concerned there was likely to be little difficulty for labourers and artisans alike, for an army of such men would have been required to clear the site of the rubble, nor was there likely to be any delay in getting reconstruction started as there were no more than three months before the autumn rains would set in.

The vicar, on learning of the outbreak soon after the morning service, had hastily thrown off his surplice and hurried to the fire in New Street. He had worked tirelessly to salvage furniture from the cottages, had gone from place to place seeking refuge for the elderly and the sick and was still busy when the clock on his church struck midnight.

The *Western Luminary* said, in its stop-press report, that the total number of dwellings and other buildings destroyed amounted to one-hundred-and seventy-four and went on: 'Business is quite at a standstill in the town and the scene of desolation is truly distressing, no object presenting itself to the eye in any direction but blackened rafters, dismantled buildings and ruins still smouldering.'

The old workhouse and the church were pressed into service to shelter the homeless and, once again, the people of Tiverton, six miles away, sent aid in the form of bread, meat and ale while Upcott's Blanket Mills in Cullompton must have supplied many with warm coverings for the night.

Less than four years later, on a blustery March morning in the year 1843, a fire broke out at the Red Lion Inn and raged with considerable fury and, despite valiant efforts made by the local fire brigade, the building and its stables were destroyed although every horse was saved. Unfortunately it proved impossible to save eleven cottages in the vicinity which, although the blaze lasted only a little above an hour, were burnt down with all the furniture they contained.

The *Exeter and Plymouth Gazette* reported: 'The whole of the property is insured in the Yorkshire and the Farmers' Offices. Owing to the exertions of the West of England's Brigade of Firemen the Half Moon Inn was preserved.'

Once again the event draws attention to the way in which such fires as this, a few cottages here, an inn or a school there, make it impossible to talk of these small Devonshire towns as belonging to any architectural period because parts were constantly being destroyed and replaced. That they are the sites of settlements of great antiquity there can be little doubt, but the actual buildings, with the exception of parish churches, Guild-halls and the like, must be difficult to date with accuracy but are generally 19th century or later.

By May of 1852 the West of England Fire Insurance Company had reached the conclusion that it would be to their advantage to keep an engine permanently at Cullompton and, in the issue of the *Exeter and Plymouth Gazette* dated 22nd May of that year, the reporter states: 'Cullompton. A fire broke out yesterday (Thursday afternoon) at the shop of Mr. Turton, the baker. The roof being thatched it spread with great rapidity to the whole of the adjoining property, it being thatched also, and fears were entertained that the conflagration would prove extensive. The West of England Fire Brigade with their engine was quickly on the spot and it was entirely owing to their exertions that the adjacent property was saved.'

Obviously the engine had not come from Exeter as it had done in the case of the 1839 blaze, and it is an indication of the co-operation then existing between the various insurance companies that the report was able to add: 'Mr. Turton was insured in the Star Fire office.'

The introduction of this third fire-engine to the town together with the rebuilding of some of the burnt-out houses with brick and slate seems to have decreased the number of fires, and what the *Exeter and Plymouth Gazette* had once called 'The scourge of the town' promised to be a thing of the past. However, Cullompton was one of those small towns with groups of cottages here and there around its outskirts, many of which were very old and, all too frequently, in a state of disrepair. In just such a row of old dwellings a fire was accidentally started by a child who lit some shavings on the hearth to dry his boots.

The flames set fire to surrounding woodwork and in a moment the place was well alight. This occurred on 9th January 1863, in the cottage of a stonemason called John Beazley, between one and two o'clock in the afternoon, when all the men and many of the women would have been away at work.

The boy who started the fire was the mason's nephew. He was about eight years old. All the dwellings in the group were thatched and were quickly alight; in the stiff breeze blowing at the time, they stood no chance against the flames. The timbers were very old and tinder-dry so that in less than an hour the entire row was a heap of ruins.

Three of the cottages were unoccupied. They were the usual type of labourers' cottages, having one room on the ground floor and one bedroom above, and it is interesting to know who, and how many, occupied each. The stonemason, his wife and nephew occupied the first in the row. Next door lived James Routley with his wife and three children; then came Henry Beazley and his two grandchildren; next to them lived William Gillard with his wife and six children; then came William Hart with wife and three children and finally Robert Kerslake, who lived alone. Thus in under an hour twenty-five people were made homeless and all their possessions destroyed.

The fire-engine of the Royal Farmers' Insurance Company and that of the West of England attended this blaze and the pat on the back given them by the *Exeter and Plymouth Gazette* seems a trifle overdone: '. . . did great service in staying the progress of the flames'. It is difficult to see how this remark can be justified in view of the absolute destruction of the entire row.

The gutted cottages belonged to a Mr. W. Frost, chemist, whose shop was near the town centre on the corner of Church Lane. He was, the newspaper assured its readers, insured in the Sun office for £220. 'Which however will not cover his loss,' the report added.

A little more than five years later the Cullompton 'scourge' again struck in no uncertain manner. The *Exeter and Plymouth Gazette* of 1st July, 1868, introduces a fire report with the general observation which reads: 'At this season of the year, after a lengthy period of hot weather, we have generally to record a series of conflagrations in the towns and villages of Devonshire which have not yet progressed beyond the stage of thatched houses. The description of these fires is nearly always the same. From some cause or other one house is set in a blaze and the wind scatters the fiery particles abroad. The dry thatch of neighbouring dwellings is soon in flames and an entire block or row of houses speedily becomes a mass of blackened ruins.' It was so in Cullompton early on Saturday morning, that being the 24th July of that year.

Shortly before three o'clock in the morning the town's postman, Mr. Reynolds, set out along the silent, deserted main street on his way to the railway station to collect the overnight mail-bags. As he was passing the chemist's shop of Mr. W. Frost he observed a flickering light through the shutters and the fanlight over the shop door. The shop was on the corner of Fore Street and Church Street and was the property and home of the owner of the eight cottages burnt down in 1863.

Running into Church Street the postman discovered that the rear of the premises was in flames. Immediately he ran across the street and roused Mr. Isaac Frost, the officer in charge of the West of England fire-engine, the Royal Farmers', and the two town engines. Not a moment was lost in summoning the firemen and getting the engines working on the blaze. The chemist and his family were quickly roused before the postman went on his way to spread the alarm through the town.

The startled inhabitants were soon on the spot while those living behind the chemist's shop were not a moment too soon in getting out. The roof of Mr. Frost's shop was slated but those houses adjoining, both along Fore Street and down Church Street, were thatched.

The fire had started in an untenanted house at the rear of Mr. Frost's shop, which was used by him as a store for oil and other inflammable substances in quite considerable quantities. From this oil and the spirits flames shot up, setting fire to the thatch of all the closest houses, and licked up the walls of the slated building. They set fire to the rafters and did the same to another house of similar construction, the home of a cab proprietor, and so fierce was the heat that it appeared impossible that it could be quelled before it burnt itself out. In the yard behind the cabman's house were a number of thatched outbuildings, including the stables but all these, through the untiring efforts of neighbours, were saved by being constantly doused with water thrown from buckets.

The row of houses next to the chemist's in Fore Street had all their roofs in a blaze at once. Quickly as the fire-engines were manhandled into position, ten houses were already burning before the first jet of water could be directed on to them.

The water seemed powerless to quench the flames and the entire row was soon a glowing mass so hot that the hosed water appeared to turn to steam before even reaching the flames. An onlooker described the houses as 'belching forth fire like a volcano'.

The hopelessness of trying to save the ten dwellings was quickly realized and the engines drew off and were strategically placed so as to prevent the fire spreading beyond them, and by dint of great exertion the ruin was confined to eleven houses and not before several large properties suffered roof damage.

Two cottages below the Crown and Thistle Inn, some fifty yards away, were badly damaged but not destroyed. After six hours the fire was got under control but the widely held illusion that houses with slated roofs were safe from danger was shattered. The chemist's shop was reduced to a heap of cinders before the onlookers' eyes and it was deemed necessary for the engines to play their hoses on the gutted buildings all that day. As late as eight o'clock in the evening the charred embers were still smoking and even occasional flickers of flame could be seen.

Mr. Frost's shop and all his back buildings were destroyed together with all the stock-in-trade. Next door was the cabinet-maker, Mr. Gillham, who also lost everything. Mr. Lenton, basket-maker, lost stock and tools as well as his house and its contents. Mrs. Harvey, the tea dealer, also lost everything she possessed, including a large quantity of tea. Another of the ruined houses in the row had been occupied by Mr. Harris, the perfumier, who, unlike all the other occupiers, had neglected to insure against fire. From the eleven houses nothing was salvaged except what the tenants could carry by hand, and as the fire started so early in the morning few of them were even properly clothed.

The *Exeter and Plymouth Gazette*, while conceding that slate-roofed houses could be destroyed by fire, reminded its readers how many such buildings were preserved when those under thatch were burnt out, and went on: 'although thatched roofs are so dangerous several houses (in Cullompton) have had new thatches put on and the National School is being recovered with straw at this very time. It takes a good deal to warn some people. There had not been so great a fire in Collumpton since the disastrous calamity of 25 years back.' The report goes on to estimate the cost of the damage at 'about £5,000'. How this fire started remains a mystery.

9

STOKE CANON

The small village of Stoke Canon, about four miles to the north of Exeter, stands close to the spot where the river Culm runs into the Exe. Before the middle of the 19th century the population of the entire parish amounted to less than five hundred persons, accommodated in about one hundred houses, many of which were scattered around the village at a distance of anything up to a mile or more. The actual village, therefore, comprised probably no more than fifty or sixty dwellings.

Two pointers suggest that Stoke Canon was a prosperous place: first, that the parish church supported a living of £221 per annum and second, that the sum of £1,000 was spent on rebuilding the Church in 1836, after which only the tower remained of the original structure; to this was added a new clock and four bells.

Even the better types of dwellings were roofed with the dangerous straw thatch and, on 10th May, in the year 1841, at about eleven o'clock in the morning, a fire broke out in a cottage close to the Stoke Canon Inn. The blaze quickly spread to other roofs nearby, including that of the inn.

No time was lost in sending a messenger to Exeter for help and, with commendable speed The Little West and the fire-engines of the Norwich Union and the Sun insurance companies were soon fighting the outbreak. In spite of a good supply of water and the indefatigable brigades' efforts the fire could not be subdued until after four o'clock in the afternoon, by which time no fewer than twelve dwellings, including the Stoke Canon Inn, were completely destroyed. No lives were lost but damage to furniture was considerable, as what was not actually burnt was badly damaged in the haste to retrieve it from the blazing houses.

The *Exeter and Plymouth Gazette* of 15th May, reported: 'How the fire originated seems to be a matter of much doubt', a remark which savours of the 'newshound' sniffing for the sensational, and it continued, 'but it is very fortunate it did not occur in the night or inconceivable mischief might have resulted from it,' at which point the reporter's financial instincts prompted him to add: 'most of the property, we understand, is insured and a subscription has been entered into for the relief of the poor sufferers.'

Here again is a hint of prosperity among the local landlords who, in rebuilding the gutted houses and the inn, replaced the old cob and thatch with brick and slate as became evident when, six years later, an even more disastrous conflagration occurred.

At about half past three o'clock on the afternoon of Tuesday, 27th April, the down train from Paddington, on the recently opened line to Bristol, which passes close to the village, was slowing as it approached St. David's Station, Exeter. As it drew level with some thatched buildings a strong gust of wind caught some sparks from the firebox and scattered them on to the roofs which instantly flared up. The building so ignited was an unimportant linhay behind the house of a Mr. Hunt, a small farmer who also kept a butcher's shop. He was, at the time, in bed ill. The gusting wind quickly fanned the fire on the thinly thatched linhay roof until every strand of straw was burnt. Before the rafters could catch fire, however, much of the flaming debris landed on the roof of the dwelling-house, which was soon ablaze, but the sick man had been carried to safety before the roof fell in on to his bed.

From Mr. Hunt's house the fire spread rapidly in two directions, easterly towards the main street and northerly towards the Brampford Speke road. On a corner of this road stood a large house, occupied by a Mr. Greenslade, a cattle-dealer. The fire raged on either side of the house for more than an hour, its thatched roof several times catching fire, but men on ladders beat out the flames with brooms and splashed buckets of water over it until the house eventually escaped with only slight damage.

Passing by Mr. Greenslade's house the flames veered off diagonally to set fire to the roofs and burn down the whole side of the street which leads to the railway crossing. Leaping across the main road the flames ignited the roof of the New Inn, a row of cottages and a tythe-barn, part of which was used for a school and part as tenements.

The nearby substantial house of Mr. Haycroft, a fellmonger, was gutted and among his losses were one hundred packs, or bales, of wool. Nor did the raging fire show any reverence for the clergy but quickly set fire to outhouses belonging to the Rev. J. Bond who was, however, absent from home. The stable containing several horses and much tack was threatened but the exertions of neighbours prevented its destruction and saved the animals. On the opposite side of the street, meanwhile, every effort to save the King's Arms failed and the inn was gutted.

From here the flames spread to the house of the vicar, the Rev. W. T. Napleton-Penoyre, an absentee whose vicarage was then occupied by the Rev. T. H. Knight, who happened to be on visit to his surgeon in London at the time. By the greatest misfortune the fire at the vicarage was not discovered until the house was well alight and very little of its furniture or the personal belongings of Mr. Knight could be saved, but a brave member of the West of England fire brigade succeeded in saving some of the church vestments and plate, although most was destroyed.

The *Western Luminary* concluded its report on the fire thus: 'We can give no accurate estimate of the amount of damage sustained but it must have amounted in the whole to many thousands of pounds. At least 30 houses including the Vicarage house, both Public-houses, the Post Office and the residence of Miss Craig, who keeps a highly respectable Ladies Boarding School, were entirely destroyed and the greater portion of their contents perished in the flames.' Stoke Canon presented a grim spectacle that evening, with very few of the thatched houses remaining standing, while all the recently built brick-and-slate dwellings had escaped.

Those people of the village who had been fortunate enough to save any of their possessions took them for safety to the church and many sheltered there themselves. They were destined to live and sleep there for many months.

A circumstance which had contributed to the ruin of the village was the absence from Exeter of most of the city's fire-engines, which had been called to fight a blaze at Exwick, a mile and a half in the opposite direction. Nevertheless, The Little West contrived to reach Stoke Canon in little more than half an hour of being summoned and the Sun engine was only ten minutes behind. Another engine pressed into service that day was the privately owned one belonging to Sir Thomas Acland at Killerton. The strong wind carried clumps of thatch as large as dinner plates through the air, keeping the inhabitants constantly on the alert for their own personal safety. From the first minutes of the outbreak as many as four separate fires were burning at one and the same time so that the people hardly knew where their services were most needed. On several occasions the fire-engines were short of water because the boards used for ponding it back as it ran off the burning buildings had been burnt early in the proceedings. There was plenty of water available at about a mile distant from the main blaze and valuable time was lost in organising ways and means of fetching it.

In such a comparatively small community it was natural that villagers should look to their wealthy landlords for immediate help. Lady Acland and the Rev. J. Bond, together with the gentry and farmers of the neighbourhood, were quick to assist the sufferers. It was necessary to start a subscription to replace furniture and tools and this was immediately organized and widely advertised. It was at first feared that a recent public subscription for famine relief in Ireland might jeopardise Stoke Canon's chance of a liberal response, but this did not prove to be the case. Furthermore, after some careful enquiries had been concluded, the Great Western Railway Company paid half the calculated cost of the uninsured loss, which totalled about £10,000. It was perhaps appropriate that the Railway Company should have accepted at least a substantial part of the cost as this incident may have been among the first in which a locomotive was found to have been directly responsible for a serious conflagration.

The *Western Luminary* reporter waxed eloquent in his description of Stoke Canon at the end of that fateful afternoon. 'After the fire,' he wrote, 'the village presented a dismal and distressing appearance.' Where before pretty cottages with neat gardens had adorned the prospect and entertained the eye, nothing was to be seen but tumbling, blackened walls, gaunt chimneys rocking in the wind, charred rafters and door-posts and smouldering ruins, but this melancholy metamorphosis conveyed nothing of the ruin which had settled on the unfortunate inhabitants, compared with the scene presented by the interior of the village church, where the burnt-out population had taken up their abode during the tempestuous night. 'Every part of the sacred edifice excepting only the area within the Communion Rails,' he noted with evident approval, 'as also the pulpit and reading desk were occupied by fragments of household furniture.' Then the reporter turned to take a closer look at the people and went on: 'Infants were cradled in the Font and in the Vestry an aged female sat alone in her desolation, surrounded by the chattels she had saved from the wreck of her fireside. In the large square pews of the Church, beds had been made up and it appeared that the darkness of the place had been rendered visible by flickering candles during the silent watches of the night.'

Stoke Canon being so near to Exeter, it was not surprising that sightseers crowded the road to the village. Upon discovering that the only two public houses had been completely destroyed, many walked a further mile to the sister village of Rewe, only to learn that its only inn had been burnt down a year before and not rebuilt.

In a little more than a month, that is to say on 3rd June 1847, Stoke Canon suffered another fire when the property of Mr. R. Chamberlain, a butcher and small farmer, which had escaped with minor damage in the previous outbreak, was set alight accidentally. This time the famous Little West was not engaged elsewhere and was quickly on hand, together with other engines, and between them they were able to confine the blaze to the premises of Mr. Chamberlain, the whole of which were gutted.

At four o'clock in the morning several members of the butcher's family started their day by preparing breakfast and had lit a fire with dry reeds and some furze. Unfortunately an opened door created a draught and a mass of sparks was carried up the chimney on to the thatched roof of the house. The weather had been extraordinarily hot and dry for some weeks past so that the thatch was like touchwood and it was instantly in a blaze which swept through the house and outbuildings, enveloping all in flames.

Those members of the family who were still in their beds included Mrs. Chamberlain, who had a narrow escape, reaching safety wearing only her night attire. So hurriedly did she run from the bedroom that she forgot a bag containing £64 which lay on a small table. On reaching the street she remembered the money and began to bemoan her loss. Instantly one of her daughters, who had risen earlier and was fully clothed, ran back into the smoke-filled house in an attempt to reach her mother's bedroom. The girl somehow succeeded but whilst going down the already burning staircase it seems probable that her apron caught fire. She had the sense to throw it off and it fell into the flames with her own purse, which contained two sovereigns.

The destruction of the entire property was so complete that nothing was saved other than that which members of the family carried upon their persons. The entire contents of the house together with all the live and dead stock in the outhouses was burnt to a cinder, even the carcase of a large, valuable bullock of very high quality which had only been killed the night before for hanging in readiness for the Exeter Friday Market. A sheep, two lambs, four calves, thirteen pigs and probably more than one hundred fowls were also destroyed. In attempting to save some part of this property Mr. Chamberlain sustained very severe injuries requiring the medical care of a surgeon. It was as well that his daughter saved the mother's cash as the amount for which her father was insured did not cover his losses.

It is difficult from this distance in time to look back on the scene and imagine what the members of such a comparatively prosperous family must have suffered as they watched helplessly whilst their home and business were comprehensively wiped out in so short a time.

The newspaper report tells us no more than the bald facts in this instance and can only tantalize the curious as to the consequences. Somehow the fire of a few weeks earlier evokes less speculation about the sufferers, perhaps because there were so many to share their misfortune while, in the case of the Chamberlains, the calamity had struck just one family. The village, too, was largely still in ruins, the very sight of which must have added to their depression, and their hope of finding any accommodation in the remaining habitable houses must have been small indeed.

10

SOUTH MOLTON

During the 19th century South Molton was one of the larger Devon market towns and had, at its busiest period, a population in excess of 3,000 persons. It stands in the centre of north Devon on the west bank of the River Mole and its burgesses, unlike those in many other towns, showed themselves to be alert to the need for fire precautions.

As far back as the year 1736 there is a record in the annals of South Molton showing that a large four-wheeled fire-engine was purchased with public funds and more than fifty years before that a mayor had purchased a number of leather buckets which were hung in the Guild-hall for use by anyone in the event of fire breaking out in the town.

With such a history of civic caution it is a little surprising that local property-owners were not persuaded to build in less inflammable materials than the common cob and straw thatch, which might have reduced considerably the number of outbreaks which occurred in the town.

The design, or plan, of the town was simple. In the centre was a square from which the streets radiated, the longest, running from east to west and about three-quarters of a mile in length, was called East Street. South Street, which runs laterally at the foot of a slope, then included a number of houses known as Causeway or Causey, while, at the further end, and forming another side of a rectangle, was North Street. All the important public buildings were situated in the square and comprised the Guild-hall, the covered market with the assembly rooms and the post office, which had once been the corn market.

On June 22nd in the year 1825 *Trewman's Exeter Flying Post* reported a fire which broke out at a blacksmith's forge in East Street. It happened between the hours of ten and eleven o'clock at night and quickly spread to twelve small cottages in the immediate neighbourhood. It was midsummer and little rain had fallen for several weeks so the straw thatch of the roofs was easily ignited. So rapidly did the fire spread that the occupants of the cottages had no time to save more than a few pieces of their furniture and had not the inhabitants of the town turned out in force to help the moment the alarm was raised, and the fire-engine been so rapidly deployed, the flames might have engulfed many more dwellings and outhouses.

At midnight on 14th March 1826 a carpenter's shop caught fire. It was close to the turnpike gate on the Exeter Road. The fire destroyed the toll-house and four other houses close by before the fire-engine's crew and helpers could put out the flames. Again there was no loss of life but a great deal of household furniture was broken or burnt. A traveller who was passing the gate spotted flames in the carpenter's workshop and raised the alarm. A newspaper report hinted at carelessness in leaving wood shavings too close to a fire whose embers had not been properly dowsed before being left for the night.

In high summer in the following year, 1827, the house of a local surgeon, a Mr. Flexman, was burnt down, together with part of an inn. It was soon after noon on Saturday, 14th July, and the fire was thought to have started in the back premises of the Ring of Bells, which was badly damaged. The stables and several newly built tenements were also gutted. No horses were in the stables at the time nor was anyone among the occupiers of the tenements injured. In this case, sparks from a kitchen chimney were thought to have set fire to the stable roof and the people from the tenements had time to remove all their furniture to safety.

Sparks from kitchen hearths or bakery ovens were usually found to be the cause of the outbreaks but every now and then someone bearing a real or imaginary grudge against a property-owner would deliberately start a fire on his premises. If the arsonist was content to stealthily light a single fire in the thatched roof or hayloft it was almost impossible to detect the crime, but suspicion became certainty when the culprit was foolish enough to kindle several fires, on one man's premises, at the same time.

In 1833 there must still have been a small amount of woollen manufacture carried on in South Molton. Evidence of this appeared in the *Exeter Flying Post* of 18th July, which reported a fire that occurred on Friday, 12th July: 'Some warehouses, a barn, stables and a large rick of wool, together with a considerable quantity of wool, the property of a Mr. Hitchcock, Woollen Manufacturer, were destroyed.'

The blaze had started about an hour before midnight and, it was suspected, had been started deliberately. In this instance no dwellings were destroyed, no lives lost and the newspapers kept their coverage to a minimum, but the reference to the wool destroyed is interesting because, during the following decade, the closing of all the mills in the town led to a reduction of as much as twenty per cent in the population. This loss by fire may well have precipitated the decline of the local trade.

In the following week's issue of the same paper a report states: 'There has been another fire at South Molton very near the spot where one occurred last week and on the property of the same individual, Mr. Hitchcock, who gives employment to at least a hundred persons who would otherwise be unemployed. Very little doubt exists,' the paper went on, 'that it was the work of incendiaries.' Then follows the ominous remark: 'We hear that Mr. Hitchcock intends to close his concerns in consequence, the effect of which will be to throw many persons now supported by labour under him, on the parochial funds.'

On a pleasant Sunday afternoon at the end of May, in the year 1837, a service was about to begin in the parish church when from somewhere within the building an alarm of 'Fire!' was raised. The assembled congregation instantly presumed that the church was burning and panic ensued. Men and women scrambled over the seats in a mad rush for the doors. The hour being soon after three o'clock, many other people were out taking an after-lunch stroll when the congregation suddenly erupted into the street.

Decorum seems to have been abandoned as women pulled up their skirts to free their legs to climb, jump and run and many lost shoes while others, whose hats had fallen off, were forced by pressure from behind to leave their best millinery to be trampled underfoot. Gloves and purses, prayerbooks and hymnals were strewn about the place.

After all this it was quickly discovered that the fire was not in the church at all but in the nearby workhouse, from whence streamed the ninety inmates and staff in similar confusion and disarray. They were, however, shepherded to the town hall and other convenient buildings while men, many in their Sunday-best clothes, helped tackle the blaze at the workhouse which, unfortunately, they were unable to save from destruction.

In its report of the event the *Exeter Flying Post* concludes: 'The building was insured . . . . for £300 and the fire is considered to be the work of incendiaries.' The reporter seems to have suffered from an obsession about arson as one cannot help wondering who could possibly want to set the workhouse alight.

The town of South Molton appears to have lacked none of the facilities available in larger cities, as the following opening paragraph of a fire report in the *Exeter Flying Post* suggests: '1st October, 1841. On Sunday evening at about half-past eleven o'clock a fire broke out in a house of ill fame in a part of the town of South Molton called the Causey.'

A gale was blowing at the time of the outbreak and the flames spread with furious rapidity until houses in South Street, all of which were thatched with straw, were ignited. Such was the force of the wind that entire blazing roofs were dislodged and masses of flaming straw spiralled into the night sky to fall on houses far from the original seat of the conflagration. Nearby Mill Lane was soon also in a blaze, together with many stables, linhays, barns, workshops and other outbuildings. Before the night was through, seventy houses were left gutted and about a further twenty so badly damaged as to be almost beyond repair. Some eighty-seven families, totalling four hundred souls, found themselves homeless while upwards of a further one hundred and fifty had suffered such severe damage to their homes that they, too, required rehousing for several weeks before their hastily patched-up dwellings could accommodate them again.

As was so common in these cases of widespread destruction, it was the poorest part of the town which had suffered most and few of the tenants were insured against loss by fire of furniture and effects, but the press informed its readers that of twenty-one major property-owners nineteen were fully covered by insurance on the buildings. The only casualties were three pigs which perished in their sty. An elderly man who had died the previous day was entirely cremated as he lay in his coffin, his family having only just escaped with their lives when the house caught fire.

The force of the gale that night may be judged by the newspaper report which states that: 'In consequence of the tempestuous state of the weather the Fire Engines could not be worked without great difficulty and the fire was only at last arrested by pulling down a number of the adjacent houses.' The reference to fire-engines, in the plural, shows that by 1841 the town had acquired a second appliance.

In the reports there is no suggestion that the 'house of ill fame' was the victim of incendiaries, nor even is anyone accused of carelessness: perhaps it was considered undignified to enquire too closely into the goings-on in such an establishment. Whatever the cause, it was almost a miracle that, in such a wind, the entire town was not burnt down. This must surely indicate good organization by the captain of the fire brigade and close co-operation on the part of those inhabitants who allowed their houses to be pulled down to prevent the fire spreading: not an easy decision to take.

Fires were still occurring so frequently in South Molton that in the early 1850s several of the fire insurance companies deemed it advisable to have their own engines permanently stationed in the town, as was shown by a newspaper report dated 26th August 1852, when the West of England's engine was 'almost immediately on the spot'. It is a rather curious report, somewhat more suggestive than factual, and in language sometimes bordering on the imaginative:

'The workshop of a ropemaker named Robert Sparkes was seen to be on fire at about half-past ten on the night of Monday, 23rd August, and .... burnt down under extraordinary circumstances.'

The workshop had been closed that evening an hour earlier, the newspaper said, at 6 o'clock, when a boy employed by Sparkes was sent off home after he had seen his employer lock the door.

The *Exeter Flying Post* continues: 'The alarm was given and the Fire Engines were quickly on the spot and brought into operation but the water poured forth by these did not seem to check the progress of the destructive element for a single moment.' Here the report offers no suggestion as to why the water was so ineffectual but leaves its readers to make their own speculations.

The workshop and other buildings adjoining it were well constructed of stone, with some cob, and were, of course, thatched. Beneath the workshop was stored a quantity of timber, about which the report states: '. . . this, after it caught, burnt with a fury almost inconceivable and the state of the blazing mass, when the fire was at its height, was magnificent and could be seen for miles around.'

As well as the workshop, three dwellings were badly damaged and a fourth lost its thatched roof but was otherwise only thoroughly soaked. The ropemaker, the report tells its readers, was insured against loss by fire.

About a year later the newspaper, in its issue dated 9th June 1853, reported another outbreak of fire in South Molton. It had started in the bakery of J. Vernon, soon after 11 o'clock on Saturday morning, June 4th. Flames had been observed by Dr. Riccard when riding along South Street on his way to visit a patient. Before he could raise the alarm the fire burst out with great violence.

The fire-engines of the town council, the West of England and the Royal Exchange companies were quickly in attendance, but were nevertheless too late to save the shop and house of the baker and that of his next-door neighbour, J. Paul, the grocer. No stock or furniture was saved from these two premises but the firemen concentrated their energies on trying to prevent the blaze spreading. This, however, proved beyond them and the report continues: 'While they were thus occupied and in consequence of the recent dry weather, the flames crept along the thatched houses behind with fearful rapidity and, almost before anyone was aware of it, the whole of the back premises, extending a long way down the street below the Anchor Inn, was in one general blaze'. Here the reporter really warmed to his task and went on: 'Any attempt to subdue the fire here seemed unavailing and the flames revelled in their destruction and enveloped in their fiery embraces one house after another until the scene was most terrific.'

The blaze continued for more than three hours and a number of outbuildings were destroyed as well as shops and many small cottages, whose occupiers, being away from home shopping, returned to find all their belongings consumed. Here creeps into the newspaper report an unusual note of criticism: 'A little better discipline amongst the Firemen would have led to better results perhaps and it is a great pity that the branch-pipe that serves some of the engines was not entrusted to better hands than one or two of them appeared to be. The parties referred to, instead of squirting the water (which was scarce enough) on the flames, seemed only inclined to play with each other thus trifling with destruction.' Perhaps the reporter was squirted himself and decided to show up the perpetrators of such an outrage.

Less than two years later, on 5th April 1855, the same newspaper reported another fire at the bakery of Mr. J. Vernon. This time the fire occurred at night. The alarm bell was rung and aroused the inhabitants from their beds, but the damage was small because the fire-engines were so quickly at the scene that they were able to prevent the blaze spreading and only the bakehouse and furzehouse were burnt.

The *Exeter Flying Post*, now enlarged to eight pages, stated: 'Rumours are abroad as to the origin of the fire and an enquiry on the subject is now being instituted.' But in subsequent issues of the paper there seems to be no further reference to the matter, which would suggest that neither Mr. Vernon nor anyone else was considered responsible.

On the evening of 18th February 1862 a private ball was being given in the town hall. Just as the first dance had commenced, a man rushed in and sought out the town bailiff, who happened to be one of the musicians, with a request to ring the town hall alarm bell as the church tower was on fire.

Masons at work on repairs to the south-east pinnacle had failed to extinguish some charcoal they had been burning and the bell-tower caught alight.

The fire-engines quickly turned out, but the blaze was far too high for them to reach with hoses, so that buckets of water had to be carried up the stairs to the seat of the fire, which was soon extinguished. The damage no doubt provided further employment for the masons on the morrow.

The dance at the town hall had continued uninterrupted, with all the South Molton ladies at their prettiest and would, perhaps, have provided a fit subject for the descriptive pen of someone like George Eliot.

On October 6th, the local historian, Mr. J. Mills, recorded an unusual incident. He was personally present, he wrote, at a fire in North Street which destroyed five cottages. The blaze had started some time after midnight and, at twenty-two minutes past three in the morning the town was shaken by an earthquake which was distinctly felt by everyone present at the fire.

At this period, arsonists real or imaginary seem to abound and on 30th March 1864, the *Exeter Flying Post* reported that on the previous Sunday night a cottage had caught fire in Back Lane. It was thatched and was the end one in a row and prompt action had put out the blaze before it could spread. On the following afternoon fire again broke out at the same place and, this time, burnt down the other fourteen cottages in the row and the newspaper attributed the outbreak to an incendiary, this time with some justification. It would appear that the record of the Devonshire Constabulary in apprehending fire-raisers is somewhat dismal, and those they arrested were all too often innocent, or the evidence produced against them was too inconclusive to procure a conviction.

Two quite separate fires occurred on 26th June 1867. The first, at about three o'clock in the morning, was found to be at the workshop of Mr. Balman, a currier, and it quickly burnt down a large range of similar workshops located behind his dwelling-house. Storehouses and stables were soon ablaze. Bennett, the whitesmith, also had his cottage gutted with the workshops of a basket-maker and a wood-turner, together with their entire stock-in-trade. Three fire-engines attended this outbreak where, at first, the narrowness of the lane behind Mr. Balman's house and the fierceness of the blaze, prevented them being effectively used. At length, skilful manoeuvring and determined work by the firemen and helpers alike soon began to master what had, at first, seemed a desperate situation. However, after burning for some three hours with considerable fury the blaze was brought under control.

The last smouldering embers had been put out, and the fire-fighters departed from the scene to seek a well-earned breakfast when, at about ten o'clock that morning, a second alarm was raised. An outhouse behind the residence of a Mr. Furse, a surgeon, was seen to be alight and, although the wind was only light, a number of nearby thatched cottages were soon in a blaze. These dwellings stood in a tan-yard near a large quantity of dried bark which ignited and sent sparks high into the air, this way and that, and Mr. Furse's house was also soon blazing with the rest. In addition a number of other workshops and more cottages were totally destroyed, as were the infant school and the two adjoining houses. The property of Coles, the butcher, and the Exeter Inn in South Street, the Town Arms Inn in East Street and many other properties were severely damaged, some being saved only by superhuman efforts on the part of their owners and friends.

Exeter fire-engines were requested and one from Barnstaple, and one from Chittlehampton arrived. It was at once apparent that the water supply was giving out. This was, however, remedied by carters who volunteered their services and carried barrels of water from the river.

Such was the intensity of the heat among the burning buildings that it was necessary to continue pouring water on them for the rest of that day and the five engines, with their brigades, remained on duty throughout the night, as did the special constables.

At the height of the blaze an accident occurred to a youth of eighteen years, named Whitfield. While he was pumping one of the engines stationed near the tan-yard a mass of blazing thatch fell on him and, in an attempt to extinguish his burning clothes, he jumped into a tan-pit. He was pulled out and taken on a stretcher to his home, where he died.

One method adopted by some of the insurance companies to minimise risks is revealed in a report of a blaze which took place in the town on 26th March 1864. The *Exeter and Plymouth Gazette* said in its issue of April 1st: 'About 9 o'clock on Sunday night a fire was discovered in a house opposite the National School Room, in the occupation of William Avery. It appears that Mrs Avery, on going upstairs accidentally ignited several dresses which were hung up against a wall. Assistance was immediately rendered and the flames were considered to have been extinguished but at half-past-one in the morning fire burst through the roof which was composed of thatch. Fortunately men had been appointed to watch the premises during the night or probably many lives would have been lost.'

It seems almost extravagant to have set more than one man to watch yet, despite this precaution, the house caught fire and it and nine other thatched houses were burnt down. A further five dwellings with slated roofs were quite severely damaged.

The newspaper offers some explanation for the ineffectualness of the watch put on the house, thus: '. . . in consequence of the scarcity of water and help, the houses being very old the fire was not got under control until 6 o'clock.'

On 4th October 1870, according to Mr. J. Mills's little book, another fire occurred in South Street, on the west side, and destroyed fifteen dwellings and some outbuildings. The singular feature of this outbreak was the sudden ignition, in a burst of flame, of a row of dwellings from which the occupiers had to escape naked. The spread was eventually stopped at the same spot where the great fire of 1841 had been halted. The fifteen houses were reduced to ashes in less than an hour.

Perhaps the saddest outcome of any South Molton fire was the one which occurred on Saturday, 19th November, in the year 1891, at the Union Workhouse. There had recently been introduced into one of the rooms a new lamp which burned petroleum. It was of a kind that had been responsible, since its introduction about 1885, for a considerable number of fires and many deaths from burns or suffocation. Mr. Mills, alive at the time, refers to this form of lighting as 'A lamp of the most objectionable make'.

Having just been lit, the lamp broke into two pieces as a result of the sudden heat in a cold atmosphere. The petroleum flooded a table, soaking the tablecloth before spilling on to the floor and drenching matting thereon. In a second the room was a raging inferno in which three of the workhouse inmates perished and several others were more or less severely injured.

In all about a dozen persons were in immediate peril and more could have died but for the instant arrival on the scene of Police Constable Wotton who, with the help of an inmate named John

Wise, managed to reach those in imminent danger and rescue them at considerable personal risk.

The local fire-engines were quickly on the spot and confined the blaze to the workhouse, which was burnt down.

The Royal Society for the Protection of Life from Fire presented P.C. Wotton with a certificate and five pounds in gold sovereigns while Wise received £1. The people of the town subscribed to a present for the brave constable and collected £14, with which they bought for him a marble clock, a silver watch with a gold chain and a pendant. He was also presented with a book in which all the subscribers' names were written.

Looking back to the last decade of the 19th century and recalling its comparatively slow and gentle rhythm, it seems surprising that clocks and watches that mark the passing of life should have been such desirable objects to our Victorian ancestors, yet by the time of Wotton's presentation it had become a habit for many firms to present retiring employees with a watch as a mark of esteem.

11

OTTERY ST. MARY

In the early part of the 19th century, Ottery St. Mary was still one of the more prosperous market towns in East Devonshire. Once an important centre of the serge-manufacturing trade, it had survived the industry's decline by becoming a busy market for the distribution of agricultural produce and by a limited continuation of woollen manufacture together with a thriving business in silk-making. The population in 1801 was only 2,415, which included some five adjacent villages and three hamlets. This number almost doubled by mid-century and probably accounted for the dwellings being small and closely packed together.

Like many other small towns, Ottery St. Mary had suffered a series of destructive fires over the years and had acquired three fire-engines by the time of the outbreak which occurred in Yonder Street on 3rd June 1841, when seven houses were burnt down and several others badly damaged. The close proximity of the houses to one another made it almost impossible to confine a fire to the house in which it started and on June 11th, at about two o'clock in the afternoon, another outbreak caused considerable alarm. It began in premises used by a mason, from which he also retailed beer and cider, in Mill Street.

The *Exeter Flying Post* of 17th June 1841 had an accident to report which must have had some bearing on this outbreak, in which another seven houses were destroyed. The newspaper stated: 'the Parishioners are in possession of three powerful engines which were soon on the spot but want of water at first impeded their operation,' and the report went on to say that the inhabitants of the town exerted themselves and soon organized a plentiful supply.

Mr. Sandford, Chief Engineer and Fire Officer, and a fireman named Thomas Rowland were manhandling the largest of the engines and bringing it down the main street, a quite steep incline, when Rowland fell and Sandford tripped over him.

Before either could get out of the way the engine, weighing a ton and a quarter, ran over them, leaving the engineer with a fractured thigh and the fireman with several broken ribs. They were also badly bruised and were carried to their homes, where they received attention from the local surgeons.

On Christmas Eve of 1842 this same large fire-engine attended an outbreak in Sandhill Street, but although it reached the incident a remarkably short time after the alarm had been raised, it could not prevent nine houses from being gutted. The usual shortage of water was blamed, no more than a trickle being available from a domestic pump.

Soon after noon on Friday, 19th May 1866 a woman standing at the door of her cottage saw flames shoot up from the roof of the National School. Like most of the large buildings in the town, the school had a thatched roof which was soon one mass of flames. A small cottage at the back was also seen to be burning fiercely and there was some doubt as to which building caught fire first. All the cottages in nearby Jesu Street and Yonder Street were also thatched with straw and also closely packed together, with only a narrow passage here and there between the rows. These two streets were soon blazing on both sides from end to end.

The flames went from street to street throughout the town in every direction, destroying houses large and small as well as many of the shops and outbuildings behind them.

The *Exeter Flying Post* gave a graphic description of the scene at Ottery immediately after the fire. 'It has laid at least a third of the town in ashes,' the paper said, 'and all that remains of two of the most populous streets in the West part of the town is a heap of blackened, smouldering ruins, and the streets which comprise the shops of most of the petty tradesmen, as well as the cottages of the superior class of workpeople, presents a scene which we have heard compared by a Crimes Officer, who happened to be on the spot, with the streets of Sebastopol after the bombardment.'

Row after row of small dwellings were gutted and the streets were strewn with charred furniture and burnt stock from the shops. Nothing but rubble remained where the most flourishing shops had carried on their business and it was calculated that one

OTTERY ST. MARY
by an unknown artist (*c* 1820)

Engraved by W. Angus from a Drawing by J. Bonner — *for the Beauties of England & Wales.*

CHURCH of ST. MARY OTTERY.

Devonshire.

hundred and eleven dwellings had been destroyed, together with an unknown number more or less severely damaged. This meant that upwards of five hundred persons of all ages and conditions were homeless, few of whom had managed to rescue any of their possessions.

The newspaper's estimate of the lost property's value was in the region of £25,000, but, disastrous as the conflagration was, and swiftly as it had spread, there was no loss of life.

It was at first thought that the blaze had been started by school-children playing with fire but it was soon known that they had all left the building and gone home for their dinners. The cottage behind the National School had, until the previous day, been occupied by a widow who had burned a quantity of rubbish before moving out. It is probable, therefore, that some smouldering particles had flared up and, blown by the wind, had subsequently ignited a wooden outside staircase at the school which led to the girls' classroom, from which it had set fire to the thatched roof.

The *Western Times* of 1st June 1866 states that, 'At half-past twelve the town engines arrived but within a brief space the School and seven adjoining cottages were a mass of flames which were spreading still further on a strong easterly wind, which was blowing at the time, and the thatch, having been thoroughly dried during the recent hot weather, was in a good condition for burning.'

The railway's electric telegraph system had been pressed into service and a message requesting help reached the fire insurance offices in Exeter at twenty minutes past one o'clock. By a quarter to two The Little West, commanded by Captain Newcombe, was on the road and covered the twelve-mile journey in sixty-three minutes, her crew smothered with dust and the four horses in a lather. The engines of the Norwich Union and the Royal Exchange followed a few minutes later.

The *Exeter Flying Post* drew its readers' attention to what might have been the consequences of the fire on the school stairs thus: 'This especially was as dry as tinder and the old staircase, in less than two minutes, was crackling and falling to pieces in a sheet of fire. Happily none of the children happened to be in the schoolrooms as they had, only half-an-hour before, been dismissed for their dinners. Had they been in the room it is impossible to imagine upwards of a hundred little girls scrambling over those burning stairs, without serious loss of life. In less than ten minutes the whole of the school building was enveloped in a blaze of fire.'

Some idea of the density of the buildings may be judged from the fact that, in the short distance between the school and the end of the street, there were no fewer than thirty-five dwellings, all of which were blazing in a matter of minutes. The occupants of the cottages carried out what furniture they could and either put it down in the middle of the street or took it to the houses of neighbours opposite. Hardly had they done so when the roofs of those dwellings burst into flame while the furniture left in the roadway was smashed and scattered by the arrival of the fire-engines.

The more flimsily built chimney-stacks presented a further hazard when many of them fell into the burning ruins, sending masses of red-hot debris and sparks into the smoke-filled air and causing further fires. By the time the appliances arrived from Exeter, there were a number of quite separate fires burning.

At one point it was hoped that the fire had been checked when it reached Broad Street, where the houses had slated roofs which, in spite of burning straw floating everywhere on the strong wind, did not catch light. Beyond these, however, was a cluster of thatched ones and here the fire wrought terrible havoc. 'Shop after shop with all its stock,' said the *Exeter Flying Post*, 'and cottage after cottage, with their furniture, falling a prey to the devastating flames,' and the report added: 'The action of the fire in some parts of the town was most capricious. It flitted about with every gust of wind demolishing a row of 9 or 10 houses here, 2 or 3 there, and, in skipping 50 or 60 others, passed over the road to level a group of shops or cottages.' The paper also added: 'It is noteworthy, too, that many of the houses which escaped the fire were thatched whilst many of those destroyed were brick built and slated.'

At the lower end of Mill Street stood two of the town's most important buildings, that is, from the point of view of affording employment for the workers. These were the grist mill and the silk factory. Had the former caught fire, its close proximity with the latter meant that it, too, must have been destroyed. The efforts to preserve these buildings, however, were at length rewarded with success despite the danger from a row of nearby cottages, which were gutted. Another cottage, no more than a few feet away, several times had its thatch alight but determination and plenty of water prevented its destruction.

The fire was eventually got under control at about seven o'clock that evening although small groups of houses continued to burn fiercely until after midnight, and, even on Saturday afternoon, spurts of flame would blaze up among the heaps of charred remains. At no time before the arrival of the Exeter appliances did the local engines make the least headway against the flames, largely on account of lack of water. When the insurance companies' engines arrived on the scene they had sufficient length of hose to pump directly from the river into the tanks of engines near the burning buildings, thus pouring on a constant and powerful jet.

Ottery's principal inn was the Red Lion which, with its extensive stables, was completely burnt out. None of the horses was hurt, however, although so intense was the heat from the building that for some time it was impossible to pass along the road in front of it. A number of shops and a tannery in Mill Street were destroyed with almost all their stock and the wide variety of materials burnt created strange smells and dense smoke everywhere which, in spite of the strong wind, seemed slow to clear.

At about half-past six, as the mail coach from Sidmouth passed by, the weight of the heavy vehicle shook the ground sufficiently to bring down the wall of a house and the driver and his assistant were slightly injured by flying masonry. The brilliant glare of the fire behind the wall so frightened the horses that they bolted and could not be brought under control until the coach had thundered along the narrow road for more than a mile.

The Norwich Union engine left Ottery to return to Exeter at 10 o'clock that night but the West of England remained until four o'clock next morning, moving from street to street as fires rekindled here and there among the ruins where homeless men sought any object that might not have been burnt in what, a few hours before, had been their dwellings. Long after the general conflagration had been subdued a pall of smoke hung over the town and swirled about the narrow fire-ravaged streets.

The great number of homeless made it essential that every possible place of shelter was immediately prepared to receive people and the vicar called together the local gentry and the leading tradesmen. Barns, warehouses and the only unoccupied house in the town to remain undamaged were all quickly prepared to house as many families as they would hold, while everyone whose dwellings remained intact took under their roofs as many of the homeless as could be squeezed in.

The next most important problem to be tackled was the replacement of tools of trade to those who had lost them. Immediate needs were provided from a fund set up at once by the Rev. Dr. Cornish, the vicar, to which generous subscriptions were made. Sixpence each was distributed to 266 persons on the day after the fire and a centre for hot food was set up. Later on the same day a further sum of money was distributed to heads of families to carry them on for the rest of the week.

On September 2nd, of the same year, it being a Sunday, people were coming out of church when, at a spot near the bottom of Tip Hill, they were harangued by a woman member of the sect known as Plymouth Brethren. She told those who stayed to hear her that the fire was a manifestation of God's anger against the people of Ottery St. Mary for their sinful ways. The small crowd stood facing the woman with their backs to the remains of a burnt-out cottage with its shaky chimney-stack still standing. As the preacher's voice rose in condemnation the heavy Sidmouth mail coach passed down the road and shook the ground, bringing down the chimney on top of the assembled people. Eight persons, some of them children, were killed instantly and several others died later from their injuries. The woman's name, the *Exeter Flying Post* told its readers, was Eliza Hawker.

The watchmaker's son, John Gillam, aged 7 years, was among the first victims whose mangled bodies were extricated from the rubble. The thatcher's 16-year-old daughter was killed, as was Emma Rowe, a laundress, Elizabeth Davis and a Miss Kellow, all of whom were 17 years old, and 15-year-old Emma Hake, whose father worked at the Lamb and Flag Inn. Among those who died later that night were Mary Gillam, 17, sister of the boy,

John, and a young carpenter aged 23 years. It is certain that none of these would have been standing at the fatal spot had it not been for the woman who had considered herself important enough to act as the Deity's mouthpiece.

In rebuilding the gutted houses in Ottery some improvements were no doubt made in such matters as drains and, in some few instances, by the addition of footpaths, but the general layout seems to have undergone little change and the houses still huddle together as if no special plan or design ever existed.

APPENDIX I

FIRES IN DEVON DURING THE 19th CENTURY
(excluding Exeter and Plymouth)

In addition to those which are the subject of this book, fires also occurred in the following places in Devonshire (excepting Exeter and Plymouth) during the 19th century. In each case the date given is that of the first report in *Trewman's Exeter Flying Post* and the number and letter, in brackets, indicate the page and column. Copies of these newspapers are housed in the West Country Studies Library, Exeter.

Ashburton
1827 July 5, (4c). Factory and tanyard

Axminster
1800 August 25, (3c). 20 houses in South St
1817 August 26, (4b). 4 houses
1829 October 15, (2c). 5 houses
1836 May 5, (3d). Bell Inn and 11 houses
1849 April 26, (8a). Several houses
1851 January 16, (5f). Several houses
1889 January 7, (2f). Row of houses

Barnstaple
1824 December 23, (4b). 2 houses

Bideford
1833 September 12, (3c). 4 cottages
1894 June 11, (4f). Several cottages

Blackawton
1859 December 11, (7f). 3 houses

Bovey Tracey
1888 April 5, (2b). 3 houses

Bow
1816 June 6, (4c). 5 houses
1819 March 4, (4b). 2 houses

Broadclyst
1856 August 15, (7c). 8 cottages
1864 June 15, (7f). 2 cottages
1869 August 4, (5a). 3 cottages

Broadhembury
1832 May 24, (7c). 6 houses
1856 May 24, (2c). 6 houses

Chudleigh Knighton
1832 January 26, (3c). 12 houses
1875 August 18, (7f). Chapel & 5 cottages

Clyst Honiton
1825 June 23, (1b). Most of the village

Clyst St George
1859 October 13, (5a) 4 cottages

Clyst St Lawrence
1847 August 26, (2f). 10 houses

Colaton Raleigh
1896 July 21, (4d). Village Inn & 6 houses

Colebrook
1893 August 23, (3c). 7 houses

Coleridge
1825 September 1, (4b). 4 cottages

Colyton
1834 April 24, (2f). Part of village
1853 August 18, (5c). 4 houses
1887 April 1, (3c). 5 cottages

Coombe St Nicholas
1883 May 2, (3b). Part of village

Crediton
1832 May 31, (3a). 3 houses
1835 August 27, (3c). 4 houses
1836 February 25, (3a). 14 cottages
1841 June 17, (2f). 33 houses
1842 July 28, (2f). 30 houses
1847 April 29, (3c). Inn & 3 houses
1849 February 15, (5b). 7 houses & butcher's shop
1849 July 19, (8d). 13 houses
1850 April 4, (8d). 9 houses
1850 April 11, (8d). 8 houses
1850 September 19, (5e). 4 houses & malthouse
1851 December 11, (4c). 3 houses
1859 July 21, (7e). 9 houses
1860 July 11, (5e). 5 cottages
1866 July 25, (7f). 3 cottages
1886 August 17, (4a). 4 cottages
1887 July 2, (8a). 3 houses
1890 June 1, (1f). 5 cottages
1893 September 18, (3c). 5 cottages at Shobrooke

Dalwood
1825 July 25, (4b). 12 houses

Dawlish
1893 May 30, (3f). 5 cottages at Ashcombe

Exmouth
1838 March 8, (3a). 15 cottages
1861 May 1, (9c). 6 houses
1885 November 25, (3c). Inn & 6 cottages
1888 March 5, (2e). 2 houses
1888 April 30, (4a). 3 cottages

Exwick
1860 May 9, (5c). Several cottages
1873 June 18, (5f). Several cottages

Farringdon
1864 September 28, (5f). 3 cottages

Halberton
1817 July 3, (4a). 21 cottages

Harberton
1874 July 8, (7f). 2 cottages

Hatherleigh
1850 September 5, (8c). 3 houses
1860 November 7, (5b). 5 houses

Honiton
1888 December 14, (4d). Row of cottages & Turk's Head Inn

Ide
1845 March 20, (3a). 3 cottages
1864 November 9, (7c). 8 cottages
1892 March 29, (4b). 4 cottages

Ilfracombe
1883 February 28, (7f). 2 houses

Jacobstowe
1862 July 23, (5c). Several houses

Kennford
1849 May 17, (8a). Several houses

Kenton
1806 June 19 (4b). 3 cottages

Kingsbridge
1893 December 7, (3b). 2 houses

Kingsteignton
1898 October 28, (2f). Cottages

Lyn (Lynton)
1898 January 1, (2b). 13 houses

Monkleigh
1886 March 26, (3c). 3 houses

Newton Abbot
1821 August 9, (4b). 10 houses
1831 April 7, (2d). 3 houses
1856 July 24, (7f). 4 houses
1869 March 17, (3f). 2 cottages

Paignton
1820 May 18, (4b). 2 houses

Rewe
1894 July 2, (3b). 3 cottages

Shaldon
1804 October 25, (1d). 5 houses

Sidbury
1820 July 13, (4c). 10 houses
1860 January 5, (7f). 2 cottages

Sidmouth
1866 May 30, (5e). Several houses
1887 October 8, (3e). 3 cottages
1889 September 30, (4d). 2 cottages
1893 May 10, (3g). 3 cottages
1893 May 13, (7c). 3 cottages

Silverton
1837 May 25, (2c). 4 houses
1859 November 17, (7f). 2 houses
1878 April 10, (7c). 16 houses
1893 February 11, (3d). 3 cottages

South Brent
1848 June 22, (3f). 6 houses

Starcross
1855 May 31, (3c). 3 houses
1885 November 11, (5d). 5 cottages
1889 May 20, (4b). 4 cottages

Staverton
1852 May 13, (4b). 7 cottages

Teigngrace
1899 April 29, (8d). 3 cottages

Teignmouth
1821 August 30, (4c). 2 houses
1828 April 17, (2d). 3 houses
1831 March 31, (3a). 7 houses & New Inn
1851 February 6, (8b). 3 houses & Compass Inn
1885 May 20, (7f). 4 houses

Thorverton
1816 June 27, (4b). 17 cottages
1816 November 10, (4b). 4 houses
1855 October 25, (8a). 5 houses
1890 August 23, (3e). 4 cottages
1898 June 6, (4d). Several houses

Tiverton
1816 April 25, (4b). 4 cottages St Andrews St
1819 December 9, (4b). 9 houses Broad Lane
1831 June 23, (3b). 21 houses
1833 August 29, (2f). 14 houses Frog St
1836 April 28, (3a). 5 cottages
1840 January 30, (2f). 12–14 houses
1842 November 17, (3d). House & 2 cottages
1849 February 8, (8c). 14 houses
1850 March 14, (8e). 14 houses at Westexe
1851 September 4, (8e). 10 houses
1853 February 24, (8a). 6 cottages
1855 September 20, (8a). 3 houses at Washfield
1863 February 18, (5d). 8 cottages Frog St

1863 May 27, (3f). 4 houses
1871 August 30, (7e). 4 cottages Church St
1873 March 12, (7f). 5 shops & Ship Inn

Topsham
1850 April 11, (8d). Several houses
1854 February 23, (5c). 2 cottages

Torrington
1827 September 27, (4c). 6 houses
1859 February 17, (7e). 3 houses

Totnes
1895 September 13, (4f). 2 boarding houses

Uffculme
1842 August 11, (3a). 5 houses
1864 October 5, (7d). 5 houses

Upton Pyne
1850 September 5, (8a). 4 houses
1872 April 24, (7f). 5 cottages

West Buckland
1860 May 16, (7e). 5 cottages and blacksmith's shop

Whimple
1860 April 25, (8c). 5 houses

Whitestone
1875 June 9, (7d). 3 cottages

Winkleigh
1800 July 3, (3b). 18 houses
1870 September 7, (3f). 4 houses

Witheridge
1847 August 12, (3d). 7 houses & bakery

Woodbury
1864 April 27, (7d). 5 houses
1868 July 22, (7e). 10 houses

Woodbury Salterton
1865 March 22, (3f). 2 cottages

Zeal Monachorum
1861 May 15, (7c). 4 cottages

APPENDIX II

FARMHOUSE FIRES

Bideford
1823 May 15, (4c). Pittwood's Farm

Bishop's Tawton
1836 March 24, (2d). Rattone's Farm
1836 March 31, (3b). Heandon Farm

Blackawton
1839 December 3, (2c). Slade Farm

Brampford Speke
1817 May 8, (4c). Woodrow's Farm
1817 May 8, (4c). Nicholl's Farm

Buckland East
1826 December 14, (4d). Tosset's Farm

Clyst Honiton
1820 May 4, (4b). Franklin's Farm
1825 June 2, (4b). Fairoak Farm

Colyton
1832 April 5, (2c). Great Borcombe Barton

Cornwood
1826 August 24, (4b). Haye's Farm

Dartington
1802 September 30, (4c). Farmhouse

Exminster
1836 July 14, (3a). Bonhay Farm

Frithelstock
1836 April 14, (2e). Cully Farm

Halberton
1827 May 24, (4b). Farmhouse
1836 February 25, (2f). Farmhouse

Hatherleigh
1829 June 18, (2c). Seldon Farm

Heavitree
1820 June 29, (4d). Havil's Farm

Kentisbury
1836 February 11, (2f). Crang's Farm

Knighton West
1827 December 6, (4c). Grove's Farm

Lapford
1826 November 9, (4c). Callend Farm

Littleham
1826 March 16, (4b). Westdown Farm

Manaton
1818 June 11, (4b). Noseworthy's Farm

Moretonhampstead
1836 March 3, (3b). Northmoore Farm

Newton St Cyres
1840 April 30, (3a). Court Barton Farm and cottages

Sandford
1826 August 17, (4c). Allen's Farm
1830 April 15, (2e). Tremmlet's Farm

Shobrooke
1892 February 29, (4d). Bond's Farm

Silverton
1831 May 19, (3b). Hayne's Farm
1837 June 28, (3b). Ash Farm

Tawstock
1836 March 31, (3b). Wick Farm

Tedburn St Mary
1816 October 17, (4b). Farmhouse

Thorverton
1823 April 17, (4b). Radden Court Farm

Walkhampton
1801 May 4, (4e). Farmhouse

Widecombe
1840 April 30, (3a). South Combe Farm

INDEX

Index

Index